International Conference on Solid Films and Surfaces

(ICSFS 2014)

IOP Conference Series: Materials Science and Engineering Volume 76

Rio de Janeiro, Brazil
8 – 11 September 2014

ISBN: 978-1-5108-0082-3
ISSN: 1757-8981

Printed from e-media with permission by:

Curran Associates, Inc.
57 Morehouse Lane
Red Hook, NY 12571

Some format issues inherent in the e-media version may also appear in this print version.

Printed by Curran Associates, Inc. (2015)

For permission requests, please contact the Institute of Physics
at the address below.

Institute of Physics
Dirac House, Temple Back
Bristol BS1 6BE UK

Phone: 44 1 17 929 7481
Fax: 44 1 17 920 0979

techtracking@iop.org

Additional copies of this publication are available from:

Curran Associates, Inc.
57 Morehouse Lane
Red Hook, NY 12571 USA
Phone: 845-758-0400
Fax: 845-758-2634
Email: curran@proceedings.com
Web: www.proceedings.com

TABLE OF CONTENTS

IOP Conf. Series: Materials Science and Engineering **76** (2015) 011001 doi:10.1088/1757-899X/76/1/011001

Preface: 17th International Conference on Solid Films and Surfaces (ICSFS)

C A Achete[1,2], **C M Almeida**[2], **M Cremona**[3], **M Rocca**[4], **F Stavale**[5]

[1]Federal University of Rio de Janeiro, CT, 21941-972, Rio de Janeiro, Brasil
[2]National Institute of Metrology, Av. Nossa Senhora das Graças 50, 25250-020, Duque de Caxias, Brasil
[3]Pontifical Catholic University of Rio de Janeiro, Physics Department, R. Marquês de São Vicente 225, 22451-900, Rio de Janeiro, Brasil
[4]University of Genova, Physics Department, v. Dodecaneso 33, 16146, Genova, Italy
[5]Brazilian Center for Physics Research, R. Dr. Xavier Sigaud 150, 22290-180, Rio de Janeiro, Brasil

E-mail: stavale@cbpf.br

Foreword
The 17th ICSFS took place at the wonderful city of Rio de Janeiro, Brazil from the 8th to the 11th of September, 2014. The conference focused on recent advances in controlling and characterizing the physical and chemical properties of films and surfaces, with a particular emphasis on materials for electronic, photonic and spintronic applications. In addition, themes of bio-functionalized structures and devices were strongly discussed in the ICSFS, covering interdisciplinary nano and nano-bio science and technology. The conference has promoted, in various sub-fields of materials surfaces and thin films, an excellent forum for exchange of ideas, presentation of technical achievements and discussion of future directions in the field. In this volume of the IOP Conference Series: Materials Science and Engineering we are glad to present 11 peer-reviewed ICSFS contributing papers. The cross-disciplinary nature of conference topics is clearly reflected in these Proceedings' contents. The themes discussed ranged from those close to more traditional condensed matter physics, such as semiconductor surfaces to physical chemistry related issues.

The Proceedings were organized in accordance with contributions presented at the Conference. We were glad with the presence of over 160 participants, including 24 invited and plenary talks and over 50 oral contributions. We strongly believe that these Proceedings will be useful for a wide audience of those interested in basic and applied surfaces and thin solid interfaces.

Acknowledgment
We would like to acknowledge the hard work, professional skills and efficiency of the team which oversaw the general organization, particularly of Dicom (Social Communication Division) from the National Institute of Metrology, Quality and Technology, Inmetro (Brazil). We also would like to thank all the invited speakers and session chairs for making the meeting such a great success.

IOP Conf. Series: Materials Science and Engineering **76** (2015) 011001 doi:10.1088/1757-899X/76/1/011001

The Conference was supported and sponsored by Academia Brasileira de Ciências, CNPq, CAPES, Inmetro, FEI Company and Carl-Zeiss Company. We are grateful for the commercial exhibition of several leading instrument companies that provided great financial support. We thank Altmann, Bruker, CAMECA, Edwards, Horiba, M&M Vácuo, Oxford Instruments, Renishaw, RHK Technology, and SPECS.

IOP Conf. Series: Materials Science and Engineering **76** (2015) 011002 doi:10.1088/1757-899X/76/1/011002

Peer review statement

All papers published in this volume of *IOP Conference Series: Materials Science and Engineering* have been peer reviewed through processes administered by the proceedings Editors. Reviews were conducted by expert referees to the professional and scientific standards expected of a proceedings journal published by IOP Publishing.

Spectroscopic and morphological studies on interaction between gold nanoparticle and liposome constructed with phosphatidylcholine

C Tsukada[1], T Tsuji[1], K Matsuo[2], T Nomoto[3], G Kutluk[2], M. Sawada[2], S. Ogawa[1], T Yoshida[4] and S Yagi[2,4]

[1] Graduate School of Engineering, Nagoya University, Furo-cho, Chikusa-ku, Nagoya 464-8603, Japan
[2] Synchrotron Radiation Center, Hiroshima University, Kagamiyama, Higashi-Hiroshima 739-0046, Japan
[3] Aichi Science & Technology Foundation, Minami-Yamaguchi-cho, Seto 460-0002, Japan
[4] EcoTopia Science Institute, Nagoya University, Furo-cho, Chikusa-ku, Nagoya 464-8603, Japan

E-mail: tsukada.chie@e.mbox.nagoya-u.ac.jp

Abstract. The gold nanoparticles (Au NPs) colloidal solution and the phosphatidylcholine (PC) liposome aqueous solution are fabricated by the solution plasma method and the extrusion procedure, respectively. When the Au NPs colloidal solution and the PC liposome aqueous solution are mixed, considering the TEM image, we think that the Au NPs firstly are covered with the PC molecules, which do not contribute to form the PC liposome, and subsequently the Au NPs covered with the PC adsorb on the PC liposome surface. We propose that the PC molecule adsorbs on the Au sheet surface at the methyl group of N-CH_3 and the oxygen atoms of P-O, P=O, C-O and C=O bonds, because each peak intensity of N, O and P K-edges NEXAFS spectra for the PC/Au sheet is reduced in comparison with that for the PC multilayer. Furthermore, the Au NPs covered with PC seem to be aggregated each other through the hydrophobic groups of PC on Au NPs.

1. Introduction

Gold nanoparticles (Au NPs) are paid attention to use of the hyperthermia for cancer cells [1] and the drug delivery system [2] in the medical field. However, since the activity of the Au NPs is high for the biomolecules, the investigation of biocompatibility for the NPs is important. About the biocompatibility of the NPs, many investigations have been carried out by observation for transformation of cells or animal tissues before/after providing with the NPs [3, 4]. On the other hand, the investigation by spectroscopic methods other than the elemental analysis has hardly been carried out, and the cause of the above transformation has not been revealed about the adsorption reaction between the biomolecules and the NPs. Thus, in our previous study, we have selected the reaction under water environment as one of the reaction conditions in our body (e.g. temperature, several molecules and pH), and elucidated the adsorption reaction between L-cysteine and the Au NPs under water environment by sulfur K-edge Near-Edge X-ray Absorption Fine Structure (NEXAFS) [5]. Generally, the NPs colloidal solution is fabricated by chemical reduction method with surfactant molecules. However, to clarify the adsorption reaction between the biomolecules and the NPs, we have fabricated the Au NPs colloidal solution by solution plasma method, which is a fabrication technique without using the surfactant molecules [5-7]. The L-cysteine has adsorbed on the Au NPs surface at thiol group and in addition those Au NPs have been aggregated by hydrogen bonding between the amino and carboxyl groups.

Subsequently, we focus on the influence of the Au NPs about cell membrane of our body, because we think the Au NPs interact with the cell membrane surface firstly when the NPs are injected into the body and reach at the target cell. The cell membrane is composed of mainly phospholipids and membrane proteins. The phospholipids construct the lipid bilayer by binding at the hydrophobic group each other, and form the spherical shape. Therefore, the liposome formed from the phospholipids is generally utilized as the cell membrane model [8]. Moreover, since it is reported that the most phospholipids of human cells are phosphatidylcholine (PC) [9], which structural formula is shown in figure 1, we have decided to reveal the adsorption reaction between the surfaces of the liposome formed from the PC molecules and the Au NPs. The liposome is called as "PC liposome" in this paper. However, in the case of carrying out the spectroscopic measurement for the above sample, it seems hard to obtain the signal with high Signal/Background ratio for the PC molecules which interact with the Au NPs. It is because the PC molecules exist not only in outer side of the lipid bilayer for the liposome but also in inner side. In addition, in the PC liposome aqueous solution, there is also a probability for the existence of the PC single molecules, which do not contribute to form the liposome and do not react with the Au NPs. The removal of those PC molecules is indeed difficult. Therefore, as an alternative to investigate the adsorption reaction between the PC liposome and the Au NPs, we have determined to elucidate the change of the chemical states for the "PC single molecule" before/after the adsorption reaction with the "Au sheet surface" under water environment. We think that it is easily possible to retain only the PC molecules chemisorbed on the Au surface by rinsing the Au sheet with the water after promotion of the adsorption reaction.

In addition, to apply the Au NPs as pharmaceutical products, we think that the Au NPs must indicate high biocompatibility. The Au NPs colloidal solution generally includes the surfactant molecules in the fabrication process. The molecules do not indicate biocompatibility. Although the surfactant molecules are used also to the fabrication of Au nanorod, H. Takahashi *et al.* have reported that the cell viability has increased when the surfactant molecules on Au nanorod surface have been replaced with PC molecules [10]. We have a question whether the molecular replacement has been carried out completely. Therefore, to prepare the Au NPs with high biocompatibility, we decide to use the Au NPs colloidal solution prepared by the solution plasma method and to cover the Au NPs surface with PC molecules. However, the behavior of Au NPs after covering with PC is unknown.

Thus, the purposes in this study are three points. First one is to clarify the adsorption morphology between the PC liposome and the Au NPs by Transmission Electron Microscopy (TEM). Subsequently, we investigate the functional groups of the PC molecule contributing for the adsorption reaction between the PC molecule and the Au sheet surface by means of nitrogen (N), oxygen (O) and phosphorus (P) K-edges NEXAFS. Finally, we reveal the behavior of Au NPs after covering with PC molecules by means of TEM and propose the reaction model.

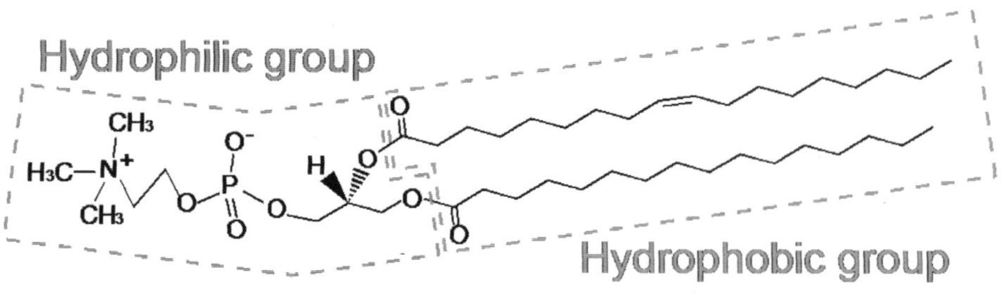

Figure 1. Structural formula of PC molecule.

2. Experimental

2.1. TEM observations

2.1.1. Au NPs colloidal solution
The Au NPs colloidal solution was prepared by the solution plasma method [5-7]. The Au wire (99.95%, 1 mm$^{\phi}$) and the sodium chloride powder (NaCl: ≥99%) were purchased from Nilaco Co. and Wako Pure Chemical Industries Ltd., respectively. The milli-Q water (≥18 MΩcm) was also provided. Two Au wires were faced each other with the distance of approximately 0.1 mm in the 10 mM NaCl aqueous solution of 180 mL. The pulse power source of AC-high voltage with both the frequency of 20 kHz and the voltage region of 3.2 kV was produced by KURITA Seisakusho Co. Ltd. When the glow discharge occurred, the Au NPs were fabricated into the NaCl aqueous solution. The discharge time was for 10 minutes.

2.1.2. Mixing PC liposome and Au NPs colloidal solution
The PC liposome aqueous solution was fabricated by means of the extrusion procedure [11], and the Mini-Extruder proposed from Avanti Polar Lipods, Inc. was employed. The PC powder (Egg, >99%) was purchased from Avanti Polar Lipods, Inc. The PC powder of 5 mg was dissolved into the milli-Q water of 1 mL and vortexed at room temperature. The PC suspension with white color was heated until 60 °C to exceed the phase-transition temperature of the used PC molecules, and went through a polycarbonate membrane with pore diameter of 400 nm to form the PC liposomes. The Au NPs colloidal solution and the PC liposome aqueous solution were mixed at volume ratio of 100:1 and vortexed at room temperature.

2.1.3. Au NPs covered with PC molecules
PC powder of 1 mg was dissolved into the milli-Q water of 2 mL. The PC suspension was kept at 4 °C and centrifuged with 20000 rcf for 5 minutes at 4 °C owing to the removal of the aggregated PC. The obtained PC supernatant solution was added to the Au NPs colloidal solution at 4 °C, and the precipitate of Au NPs occurred after several hours. The Au NPs precipitate was picked up to other micro tube and rinsed several times with the milli-Q water.

2.1.4. Conditions for TEM observation
TEM images were observed by the negative staining method. The copper grid (G200HH, Nisshin EM Corporation) with carbon/support film was provided, and the surface of the grid with film was modified to the hydrophilic state by soft plasma ion bombardment (PIB-10, Vacuum Device). Subsequently, the mixed solution of Au NPs and PC liposome or the precipitate of the Au NPs covered with PC molecules was dropped on the grid. After 5 minutes, the grid was rinsed in 2% uranyl acetate or in 2% phosphotungstic acid. In the case of use the phosphotungstic acid, the grid was subsequently rinsed in milli-Q water. The pH value of 2% phosphotungstic acid was calibrated to 7 by adding 2N NaOH. The extra staining solution on the grid was drawn off by using filter paper, and the grid was dried in air. The TEM observation under the electron accelerating voltage of 100 kV was carried out by H-7600 (Hitachi).

2.2. NEXAFS measuremenst

2.2.1. Sample preparations for NEXAFS measurements
The milli-Q water cooled at 4 °C was provided. The PC powder of 1 mg was dissolved into the milli-Q water of 1 mL and vortexed. The Au sheet surface was polished with the alumina powder and cleaned in the ethanol solution by ultrasonic cleaning method. The Au sheet was dipped into the PC suspension, and the adsorption reaction was promoted for 36 hours at 4 °C. The reaction temperature at 4 °C has been selected to inhibit the formation of the PC liposome, because it has been generally

defined that the liposome is easy to form in the case of more than the phase-transition temperature. The reacted Au sheet was taken out from the PC suspension and dried in air after rinsing with the milli-Q water. This sample was called as "PC/Au sheet" in this paper. Furthermore, the PC multilayer sample, which did not react with the Au sheet surface, was prepared by dropping PC suspension on the Au sheet and drying in air. This sample was provided as a standard and was described as "PC multilayer". As a standard for O K-edge NEXAFS spectrum, the $FePO_4 \cdot nH_2O$ powder (Wako Pure Chemical Industries Ltd.) was rubbed on indium sheet.

2.2.2. Conditions for NEXAFS measurements

To investigate the adsorption chemical states for several functional groups of the PC molecules adsorbed on the Au sheet surface, the N K- and the O K-edges NEXAFS measurements using synchrotron radiation were carried out at the BL-2 of the SR center in Ritsumeikan University [12]. The NEXAFS spectra were obtained by total electron yield (TEY) method with the sample drain current under high vacuum condition. The incident X-ray energies for the N K- and the O K-edges NEXAFS measurements were calibrated with the first peak of the NEXAFS spectra for h-BN at 399.1 eV and α-Fe_2O_3 at 529.4 eV, respectively. The P K-edge NEXAFS measurements were carried out at the BL-6N1 in Aichi Synchrotron Radiation Center (AichiSR) and the BL-3 in Hiroshima Synchrotron Radiation Center (HiSOR) [13]. The P K-edge NEXAFS spectra were obtained by fluorescence X-ray yield (FY) method using the atmospheric XAFS measurement system with He-path [14]. The silicon drift detector (SDD) was used at AichiSR BL-6N1, on the other hand, the fluorescence yield detection for HiSOR BL-3 was employed using a gas-flow type proportional counter with P-10 gas (10% CH_4 in Ar). The incident X-ray energy was calibrated on the assumption that the first peak of $FePO_4$ appears at 2153.0 eV at both beamlines. In this paper, the shown spectra of the P K-edge NEXAFS are measured at AichiSR.

3. Results and discussions

3.1. TEM observations

3.1.1. Mixing PC liposome and Au NPs colloidal solution

Figure 2 shows the TEM image of the projection view for the sample after mixing the Au NPs colloidal solution and the PC liposome aqueous solution. When we take notice of the enlarged view, it can be considered that the PC molecules adsorb on the Au NPs surface. It is because a certain bright film seems to be covered the NPs surface. The average thickness with standard deviation for the bright film is estimated to be 1.5±0.2 nm. The thickness is similar to the average length of PC molecule, which is defined as 2.0~2.5 nm [9,15]. H. Zhu *et al.* also have reported that the gold particles fabricated with phospholipids are coated with a phospholipid monolayer, because the average particle-to-particle distance is 4 nm [16] and the average distance means to be equal to the length of two phospholipids molecules. Thus, we think that the Au NPs firstly are covered with the PC molecules, which do not contribute to form the PC liposome, and subsequently the Au NPs covered with the PC adsorb on the PC liposome surface. Furthermore, considering to the results of section 3.1.2, it is speculated that the Au NPs surface is covered with two layers constructed with PC molecules as shown in figure 3 because the thickness of the certain bright film is same in comparison with the average distance between the NPs covered with PC. The PC molecules firstly adsorb on the Au NPs surface, and subsequently another PC molecule may interact with the PC molecule on the Au NPs through the hydrophobic groups.

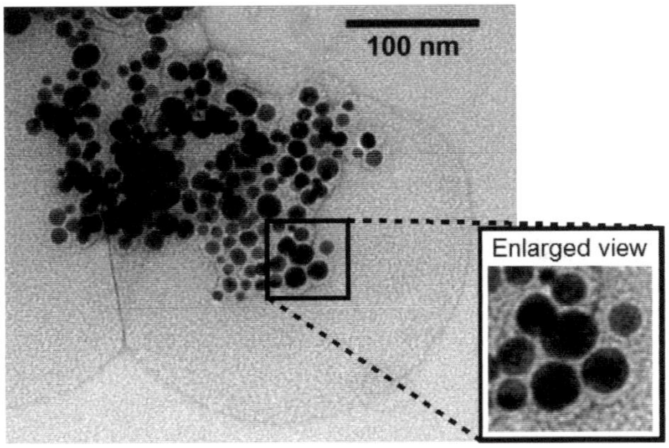

Figure 2. TEM image of the projection view for the sample after mixing the Au NPs colloidal solution and the PC liposome aqueous solution. The enlarged view is also shown.

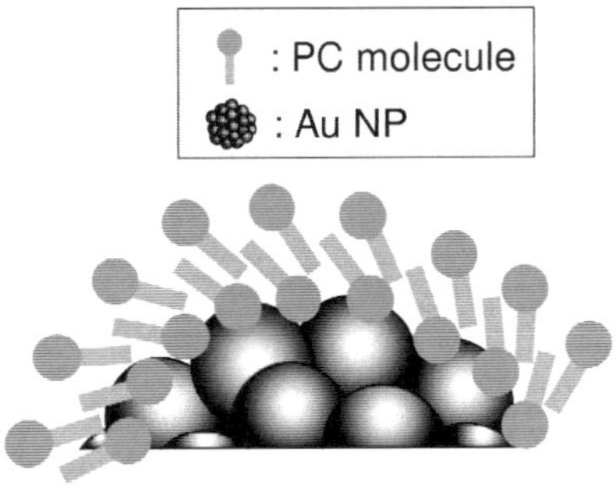

Figure 3. Schematic view for Au NPs surface covered with PC.

3.1.2. Au NPs covered with PC molecules

Figure 4 shows the TEM image for the precipitate of the Au NPs covered with PC molecules. The Au NPs are constructed like 2-dimensional beads line. In the enlarged view, the NPs are covered with bright film and the average distance between NPs with standard deviation is estimated to be 1.5±0.4 nm. Considering the results of section 3.2, the PC molecules seem to adsorb on the Au NPs surface at methyl group of N-CH_3 and the oxygen atoms of P-O, P=O, C-O and C=O bonds. Therefore, the Au NPs probably interact through the hydrophobic group of the PC molecules on the NPs. Because the average distance between NPs is different compared with the length of two PC molecules connected linearly [9,15], we speculate the possibility of three causes. First one is that the PC adsorbs on the Au NPs surface with certain inclination. Second one is that the NPs approach each other when the water in the Au NPs precipitation is dried due to preparation of TEM sample. Third one is that the hydrophobic groups interact each other at molecular side, as shown in figure 5.

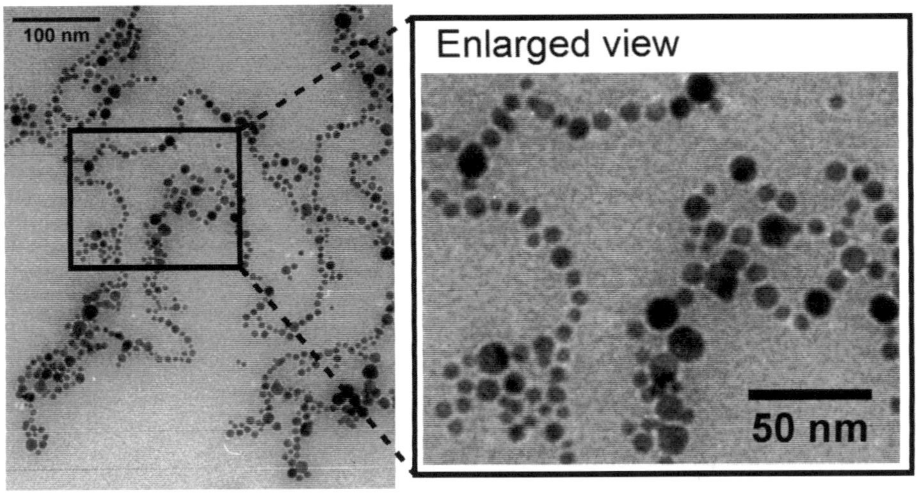

Figure 4. TEM image for the precipitate of the Au NPs covered with PC molecules. The enlarged view is also shown.

IOP Conf. Series: Materials Science and Engineering **76** (2015) 012001 doi:10.1088/1757-899X/76/1/012001

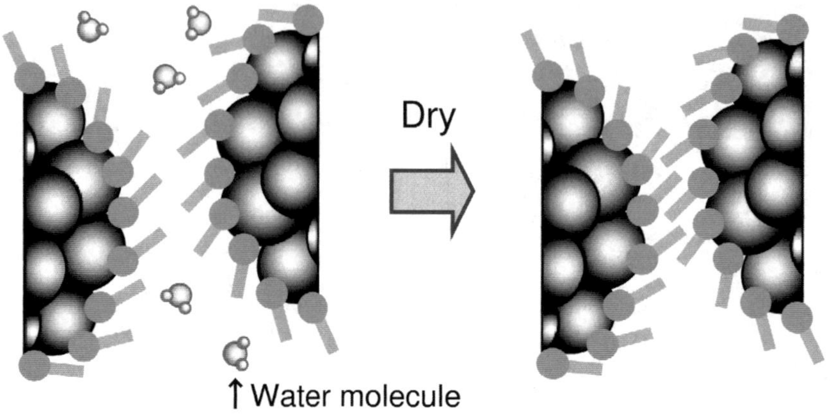

Figure 5. Schematic view for aggregation model of Au NPs covered with PC.

3.2. NEXAFS measurements

3.2.1. N K-edge NEXAFS

Figure 6 shows the N K-edge NEXAFS spectra for the PC multilayer as a standard and the PC/Au sheet. Those spectra are normalized by edge-jump. The obvious peaks for the PC multilayer and the PC/Au sheet are observed at 403.6 eV. We think that the peak position at 403.6 eV is assigned to σ*(N-C) orbital of the PC molecule. The peak intensity for the PC/Au sheet is reduced in comparison with that for the PC multilayer. S. Yagi *et al.* have reported that the peak intensity of the spectrum for L-cysteine on polycrystalline Cu surface is a little bit suppressed compared to the peak intensity for the L-cysteine bulk phase when the sulfur K-edge NEXAFS measurement has been carried out [17]. The suppression of the peak intensity has occurred by the charge transfer from the Cu surface to L-cysteine σ*(S-C) orbital. Therefore, we speculate that the reduction of the peak intensity at 403.6 eV occurs by the adsorption of the PC molecule at the methyl group of $N\text{-}CH_3$ and the charge transfer from the Au sheet surface to the σ*(N-C) orbital of the adsorbed PC molecule.

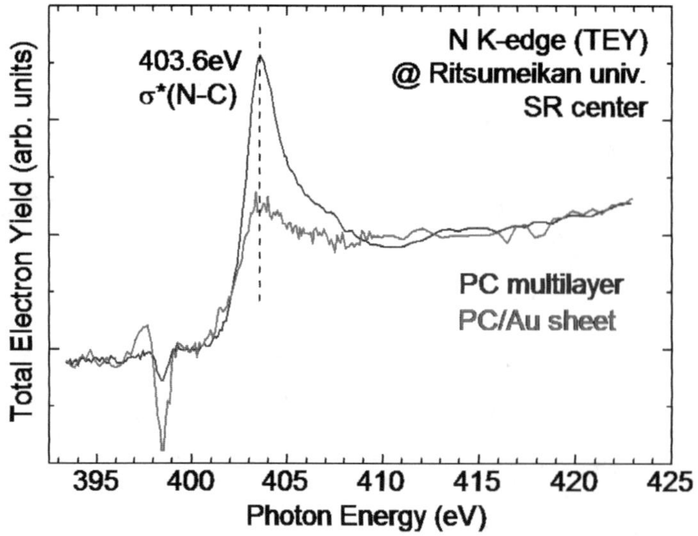

Figure 6. N K-edge NEXAFS spectra for the PC multilayer as a standard and the PC/Au sheet.

3.2.2. P K-edge NEXAFS

Figure 7 shows the P K-edge NEXAFS spectra for the PC multilayer as a standard and the "PC/Au sheet - (Au sheet)". The "PC/Au sheet - (Au sheet)" spectrum is obtained by subtracting the spectrum of the Au sheet from that of PC/Au sheet. The smoothed spectrum for the "PC/Au sheet - (Au sheet)" is also shown. The spectrum for the PC multilayer possesses a peak at 2151.8 eV, which is observed the enlargement of the full width at half maximum (FWHM) toward high energy side. Generally, if there is the double bond in a molecule and the NEXAFS measurement for the constituent element of the double bond is carried out, the π^* peak appears at lower energy side in comparison with that of σ^*. Thus, we think that the energy position at 2151.8 eV indicates the transition from P1s to π^*(P=O) molecular orbital. In addition, since the PC molecule possesses the bonds of both P=O and P-O, it is speculated that the enlargement of the FWHM toward high energy side shows the transitions to σ^*(P-O) and σ^*(P=O) orbitals. The peak and the shoulder structure for the "PC/Au sheet - (Au sheet)" are obtained at 2152.6 eV and 2151.8 eV, respectively. Because the energy position of the shoulder structure is equivalent to that of the peak for the PC multilayer and the peak intensity for the "PC/Au sheet - (Au sheet)" is reduced in comparison with that for the PC multilayer, we think that the PC molecule adsorbs on the Au sheet surface at the oxygen atom of P=O bond of PO_4 part. Since the intensity for energy region of σ^* is also reduced, it is also possible that the PC molecule adsorbs on the Au sheet surface at the oxygen atom of P-O bond. The spectra of P K-edge NEXAFS measured at HiSOR BL-3, not shown in here, show the same results.

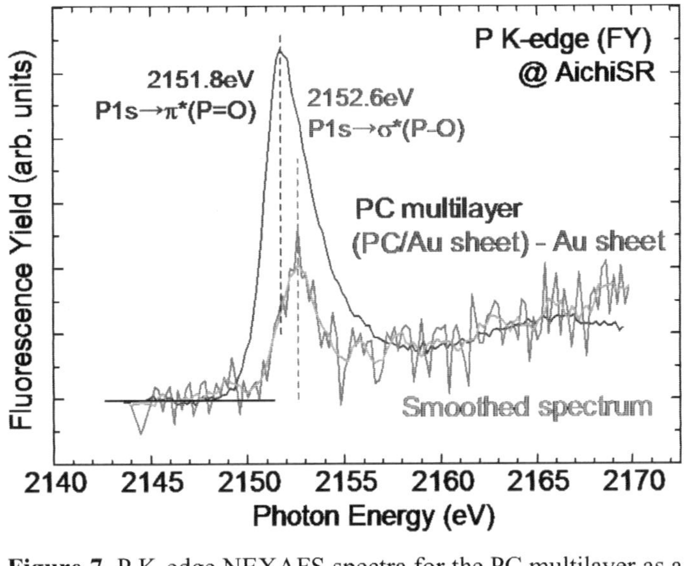

Figure 7. P K-edge NEXAFS spectra for the PC multilayer as a standard and the "PC/Au sheet - (Au sheet)". The smoothed spectrum for the "PC/Au sheet - (Au sheet)" is also shown.

3.2.3. O K-edge NEXAFS

Figure 8 shows the O K-edge NEXAFS spectra for the PC multilayer as a standard and the PC/Au sheet. The spectrum of $FePO_4 \cdot nH_2O$ powder is also shown as a standard for the unoccupied molecular orbitals of O-P bonds. This spectrum is obtained by subtracting the spectrum of the indium sheet from that of the ($FePO_4 \cdot nH_2O$ powder)/indium sheet.

The PC molecule has several kinds of O-P and O-C bonds. About the spectrum of the PC multilayer, the peak and shoulder structures are observed at 531.5 eV, 534.4 eV, 536.4 eV and around 545 eV. When we pay attention to the unoccupied orbital of O-C bond, the energy positions are derived from

O1s(O=C)→π*(O=C) [18], O1s(O-C)→π*(O=C) [18], O1s→σ*(O-CH_2) [18] and O1s→σ*(O=C) [18], respectively. Subsequently, about the spectrum for the $FePO_4{\cdot}nH_2O$ powder, the peaks and the shoulder structures are observed at 530.2 eV, 531.5 eV, 537.3 eV and 544.6 eV. We speculate that those energy positions mean the transitions of O1s(O=P)→π*(O=P), O1s(O-P)→π*(O=P), O1s→σ*(O-P) and O1s→σ*(O=P), respectively. Other materials with O-P bonds also possess the π* peak at region of 530~532 eV and σ* peak at region of 535~538 eV [19]. Therefore, we think that the spectrum of the PC multilayer is constructed with the convolution of the peaks for both O-C and O-P bonds. Although the peak at 538.3 eV for the PC multilayer is also observed, the obvious assignation of the peak has not been defined. Because the peak position at 540.0 eV shows the transition of O1s→σ*(O-C) orbital [18] and the transition of σ*(O-P) attributed to PO_4 part has also been defined at 535~538 eV [19], we think that the peak at 538.3 eV is constructed with the convolution of those peaks.
Each peak intensity of the above unoccupied orbitals is reduced when the adsorption reaction between the PC molecule and the Au sheet surface is promoted. Thus, we propose that the PC molecule adsorbs on the Au sheet surface at the oxygen atoms of O-C and O-P bonds. Especially, considering the double bonds of the PC molecule, there is possible that the oxygen atoms of O=C and O=P for the PC molecule interacts to the Au sheet surface and the double bond is dissociated. It is because the peak of π* after adsorption reaction decreases and the shoulder structures of both π* and σ* disappear. This results correspond with the results for P K-edge NEXAFS.

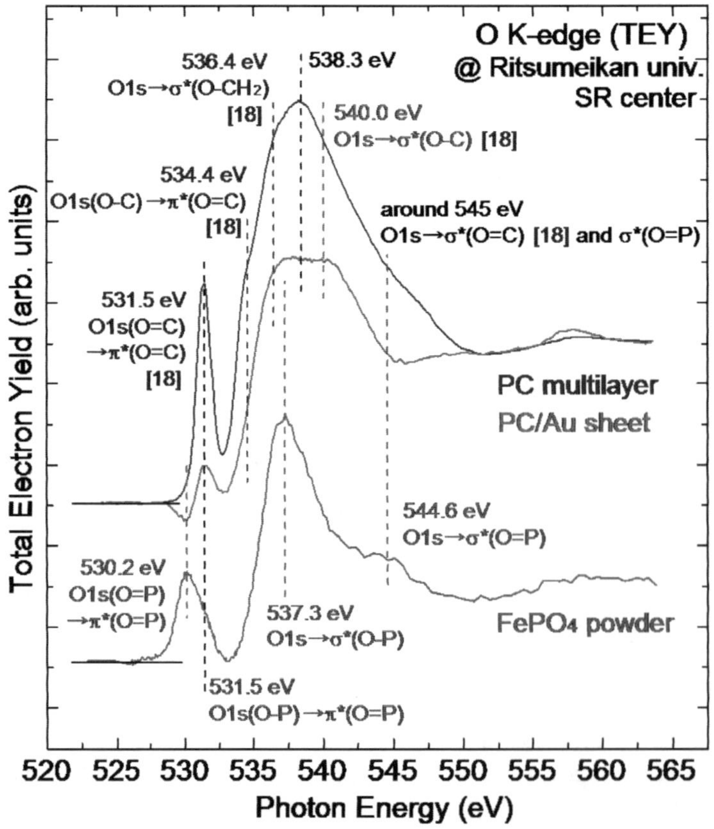

Figure 8. O K-edge NEXAFS spectra for the PC multilayer as a standard and the PC/Au sheet. The spectrum of the $FePO_4{\cdot}nH_2O$ powder is also shown as a standard.

4. Conclusions

The Au NPs have been prepared by the solution plasma method. We think that the Au NPs firstly are covered with the PC molecules and subsequently the Au NPs covered with the PC adsorb on the PC liposome surface. Because each peak intensity of N, O and P K-edges NEXAFS spectra for the PC/Au sheet is reduced in comparison with that for the PC multilayer, the PC molecule seems to adsorb on the Au sheet surface at the methyl group of N-CH_3 and the oxygen atoms of P-O, P=O, C-O and C=O bonds. The Au NPs covered with PC seem to be aggregated through the hydrophobic groups of PC on Au NPs and are also constructed like 2-dimensional beads line.

5. Acknowledgements

The authors are grateful for the financial support of JSPS Research Fellowship for Young Scientists (No. 253655). This work was performed under the approvals of Aichi Synchrotron Radiation Center (2503023 and 2503026), Ritsumeikan University SR center, MEXT project, "Open Advanced Research Facilities Initiative" and HSRC Program Advisory Committee (Nos. 13-A-26 and 13-B-24). The authors also thank Mr. Keisuke Yamanaka and Dr. Hirofumi Nameki for their helpful technical support. We thank Dr. Kentaro Fujii, Dr. Koji Nakanishi and Dr. Jiro Usukura for useful discussions.

References

[1] Huang H-C, Rege K and Heys J J 2010 *J. Am. Chem. Soc.* **4** 2892
[2] Wang F, Wang Y-C, Dou S, Xiong M-H, Sun T-M and Wang J 2011 *J. Am. Chem. Soc.* **5** 3679
[3] Cochran D B, Wattamwar P P, Wydra R, Hilt J Z, Anderson K W, Eitel R E and Dziubla T D 2013 *Biomaterials* **34** 9615
[4] Pauksch L, Hartmann S, Rohnke M, Szalay G, Alt V, Schnettler R and Lips K S 2014 *Acta Biomaterialia* **10** 439
[5] Tsukada C, Mizutani T, Ogawa S, Nomoto T, Abe Y, Nameki H, Matsuo K, Kutluk G and Yagi S 2013 *e-J. Surf. Sci. Nanotech.* **11** 18
[6] Nameki H 2010 *private communication in Aichi Prefectural Technical Report*
[7] Mizutani T, Abe Y, Hamaguchi H, Nameki H and Yagi S 2012 *Conference Proceedings NANOCON*, ISBN 978-80-87294-32-1
[8] Okumura Y, Nakaya T, Namai H and Urita K 2011 *Langmuir* **27** 3279
[9] Albert B, Bray D, Hopkin K, Johnson A, Lewis J, Raff M, Roberts K and Walter P 2004 *Essential Cell Biology,* 2nd Edition, Chapter 11
[10] Takahashi H, Niidome Y, Niidome T, Kaneko K, Kawasaki H and Yamada S 2006 *Langmuir* **22** 2
[11] MacDonald R C, MacDonald R I, Menco B Ph M, Takeshita K, Subbarao N K and Hu L-r 1991 *Biochim. Biophys. Acta.* **1061** 297
[12] Ishii H, Nakanishi K, Watanabe I, Ohta T and Kojima K 2011 *e-J. Surf. Sci. Nanorech.* **9** 416
[13] Yagi S, Kutluk G, Matsui T, Matano A, Hiraya A, Hashimoto E and Taniguchi M 2001 *Nucl. Instr. Methods* A **467-468** 723
[14] Yagi S, Matsumura Y, Soda K, Hashimoto E and Taniguchi M 2004 *Surf. Interface Anal.* **36** 1064
[15] Park S-H, Oh S-G, Mun J-Y and Han S-S 2006 *Colloids Surf. B* **48** 112
[16] Zhu H, Tao C, Zheng S and Li J 2005 *Colloids Surf. A: Physicochem. Eng. Aspects* **257-258** 411
[17] Yagi S, Matsumura K, Nakano Y, Ikenaga E, Sardar S A, Syed J A, Soda K, Hashimoto E, Tanaka K and Taniguchi M 2003 *Nucl. Instr. Methods* B **199** 244
[18] Okajima T, Hara K, Yamamoto M, Seki K 2010 *BUNSEKI KAGAKU* **59** 477
[19] Yang S, Wang D, Liang G, Yiu Y M, Wang J, Liu L, Sun X and Sham T-K 2012 *Energy Environ. Sci.* **5** 7007

IOP Conf. Series: Materials Science and Engineering **76** (2015) 012002 doi:10.1088/1757-899X/76/1/012002

Physical characterization of ultrathin silicon oxynitrides grown by Rapid Thermal Processing aiming to MOS tunnel devices

V. Christiano[1] and S. G. dos Santos Filho[1]

[1]University of São Paulo: LSI/PSI/EPUSP – São Paulo, Brazil.

E-mail: veronica@lsi.usp.br

Abstract. Oxynitrides were grown in a homemade Rapid Thermal Processor (RTP) using a low mass quartz carrier, to obtain thin oxynitrides over large areas of 3 inches silicon p-type wafers. Layers with thickness varying from 0.97 to 2.39 nm with uniformity better than 0.4%, were obtained at 700 and 850°C, in a mixed ambient of nitrogen and oxygen ($4N_2$:$3O_2$ in volume). The nitrogen concentration was obtained with the aid of X-ray photoelectron spectroscopy (XPS) and was 0.6 at%. On the other hand, the Si/O ratio in the oxynitride was approximately 1.9, indicating an almost stoichiometric SiO_2 with a small amount of nitrogen. In addition, using the $^{16}O(\alpha, \alpha)$ elastic-scattering signal at 3.039MeV, the planar concentration of oxygen was $5.5\text{x}10^{15}\text{cm}^2$ for the oxynitride grown at 850°C during 40s.

1. Introduction

The fabrication of metal-oxide-semiconductor (MOS) structures is a matter of increasing attention due to the aggressive downsizing, which also means thickness increasingly smaller. The main challenge to obtain efficient MOS tunnel devices is the achievement of high-quality uniform ultrathin dielectrics with small defect density and high dielectric strength [1-2]. Silicon oxynitrides grown by rapid thermal processing (RTP) are reported as interesting dielectrics for MOS devices, owing to their high quality of physical and electrical characteristics [1,3]. The incorporation of nitrogen during the film growth by RTP has been reported as beneficial for MOS devices due to the small amount of nitrogen distributed close to the oxynitride/silicon interface, which is responsible for a decrease of the leakage current, and an increase of the resistance to boron penetration [4].

In this work, it is shown the physical characterization of SiO_xN_y films grown in a homemade RTP using a low mass quartz carrier, considering different processing conditions (temperature, processing time and environment).

2. Experimental

Si wafers were chemically cleaned by RCA cleaning [5]. The recipe used for the RCA cleaning consisted of two baths: (a) 400 ml H_2O + 25 ml NH_4OH (38%) + 175 ml H_2O_2 (37%) and (b) 400 ml H_2O + 100 ml HCl (38%), heated at 90°C, for 15 min. Between each bath, the wafers were immersed in deionized (DI) water for 5 min. Finally, it was performed a dip in fluoridric acid, in the proportion 80:1 (80 H_2O + 1 HF(49%)), for 100 s, at room temperature and, other immersion in DI water for 3 min. Immediately after the cleaning, the wafers were oxided by RTP at the temperatures of 700 or 850°C as shown in the following.

2.1. *Homemade RTP*

The homemade RTP using a low mass quartz carrier is composed by a cover at the end of the furnace, made in quartz, with a hole to allow the quartz rod introduction. In addition, a quartz carrier for samples of 3 inches is fixed at the end of the quartz rod inside the furnace (see figure 1).

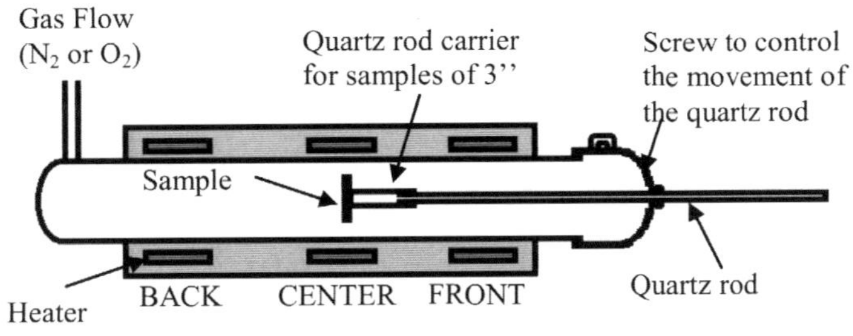

Figure 1. Schematic view of the homemade RTP.

The rapid thermal processing was performed by the quick entry of the wafer into the furnace till its center position, ensuring the natural increase of the temperature till a plateau. The quartz carrier, for 3 inches wafers, was made using three thin brackets welded on the quartz rod in order to assure low thermal mass in contact with the wafer, so that, temperature uniformity was better than 4% [7].

2.2. *Silicon oxynitrides grown*

The growth procedure to obtain the silicon oxynitrides (SiO_xN_y) were:

- The wafer was positioned in the beginning of the furnace during 5 min in ultrapure nitrogen for the wafer achieve initial temperatures of 495°C (figure 2a) or 600°C (figure 2b) for the furnace previously calibrated at 700°C or 850°C, respectively.
- The gas flow was turned on 5 s before starting the RTP. The wafers were processed in ultrapure N_2+O_2 ambient ($4N_2$:$3O_2$, in volume).
- The quartz rod was pushed in 5 s until the wafer could reach the center position of the furnace, to be processed during 10, 20, 40, 80 and 160 s, for 700 and 850°C.
- After the oxidation, the removal of the samples was during 5s, where the quartz rod was pulled until the wafer could reach the front position of the furnace and then, the oxygen flow was turned off.
- The wafer was maintained during 5 min. at the beginning of the furnace before its removal to the ambient of the clean room.

Figures 2a and 2b show temperature profiles starting with an increasing ramp, followed by the temperature plateau. The processing time is measured from the initial temperature of 495°C or 600°C till to the soak temperature of 700 or 850°C, respectively. In addition, it is important to point out that it takes 6 and 9s to achieve ninety percent of the soak temperature, respectively, for 700 or 850°C.

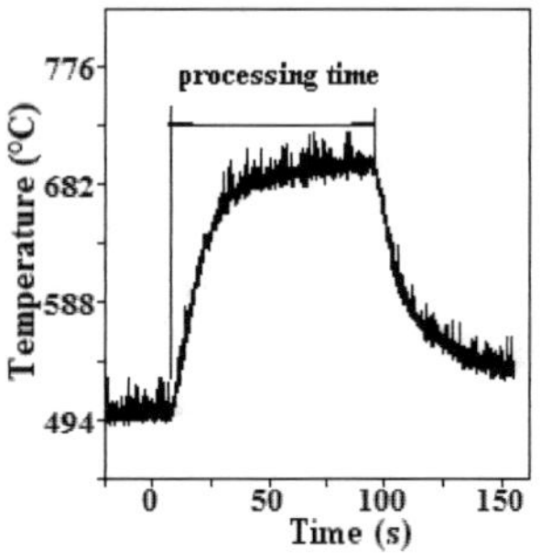

Figure 2a. Temperature profile for 700°C

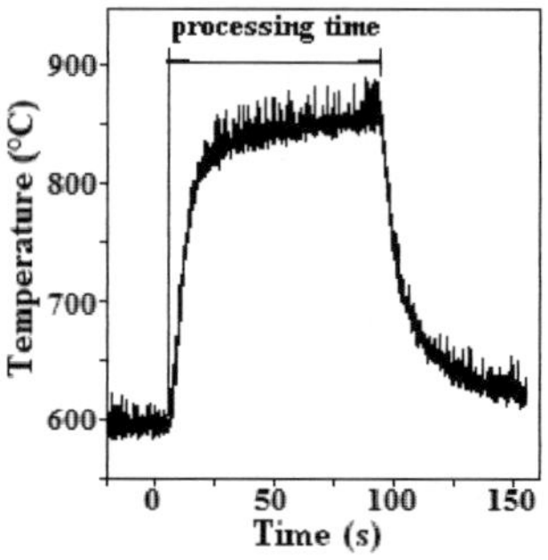

Figure 2b. Temperature profile for 850°C

2.3. Physical Characterization

The silicon oxynitride thicknesses were estimated by an ellipsometer (Autoel IV), using a wavelength of 6.328 nm. All the values of thickness in the results and discussion correspond to the arithmetic average of nine points spaced 2 cm from each other, along of the three inches wafers.

X-ray photoelectron spectroscopy (XPS) was employed to obtain the nitrogen concentration in the thin silicon oxynitrides by means of a UNI-SPECS UHV commercial spectrometer with an Mg X-ray source. The composition of the surface was determined considering the relative ratios of the peaks for Si 2p, O 1s and N 1s in the spectra, taking into account the atomic sensitivity factors [6].

RBS spectra were taken under normal incidence of a He^+ beam and with a scattering angle of 80° using an accelerator Pelletron - Tanden, model 55 DH/NEC. The $^{16}O(\alpha, \alpha)$ elastic-scattering signal was observed at 3.039 MeV, which presents to an enhanced sensitivity for oxygen detection (also known as oxygen resonance [7]). The planar concentration of oxygen was extracted from $^{16}O(\alpha, \alpha)$ elastic-scattering signal at 3.039MeV from the relation of areas of the oxygen signal for a given measured sample and a reference sample with a well-known planar concentration of oxygen, as follows [7]:

$$N_O = \left[tan\left(\frac{\frac{A_O}{A_R}.\Omega Q_{RBS}\frac{\sigma_R\Gamma}{[\varepsilon_0]^R}.tan^{-1}\left(\frac{N^R[\varepsilon_0]^R}{\Gamma}\right)}{\Omega Q_{RBS}\frac{\sigma_R\Gamma}{[\varepsilon_0]^R}} \right) \right] \cdot \frac{\Gamma}{[\varepsilon_0]^O} \quad (1)$$

Where A_O is the area of the oxygen peak of the measured sample, A_R is the area under the oxygen peak of the reference sample, Ω is the solid angle for measurement, Q_{RBS} is the particle flux or the integrated charge of the backscattered He^+ beam from the sample, σ_R is the cross section at the resonance energy, Γ is the half-width of the oxygen peak, $[\varepsilon_o]^O$ is the stopping cross section of the incident beam at the resonance energy for the measured sample and $[\varepsilon_o]^R$ is the stopping cross section of the incident beam at the resonance energy for the reference sample.

3. Results and discussions

At first, oxynitrides were grown at 700 and 850°C in a mix of pure N_2 and O_2 (4:3). The processing times were 10, 20, 40, 80 and 160 s. The graphs of thickness as a function of the processing time are shown in figure 3.

The samples processed at 700°C exhibited a rapid initial oxidation starting from the processing time of 10 s with a growth rate of 0.12 nm/s and becomes approximately constant and equal to 0.0025 nm/s in the range of 40 to 160 s. This last situation can be modeled by an oxynitridation process dominated by surface reaction according to the model of Deal and Grove [8]. On the other hand, the rapid initial oxidation can be formerly associated to an enhanced rate due to the increasing ramp of the temperature profile since the temperature profile is not enough to saturate at a plateau.

For samples processed at 850°C, an approximately linear growth rate of 0.08 nm/s was observed in the range of 10 to 80s followed by a tendency of being quadratic for processing times higher than 80 s. In this case, the rapid initial oxidation must be occurred for the processing time lower than 10 s and the growth rate became constant in the range of 10 to 80 s where the oxynitridation process is possibly dominated by surface reaction according to the model of Deal and Grove [8].

In addition, the oxynitrides films grown with thickness in the range of 0.97 to 2.39nm, as shown in figure 3, presented uniformity better than 0.4% in large areas of 3 inches silicon wafers.

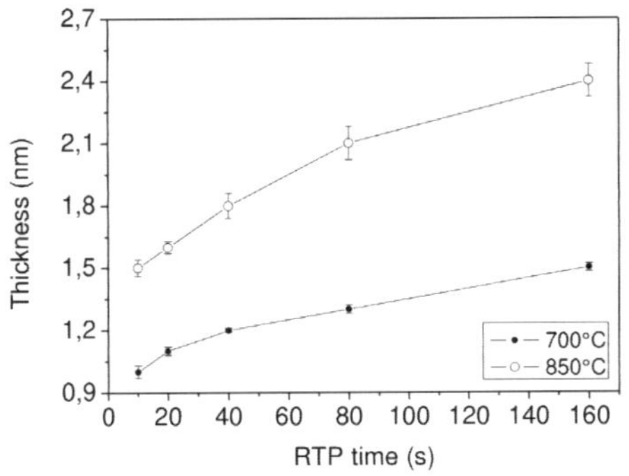

Figure 3. Average thickness as function of the RTP time for 700 and 850°C.

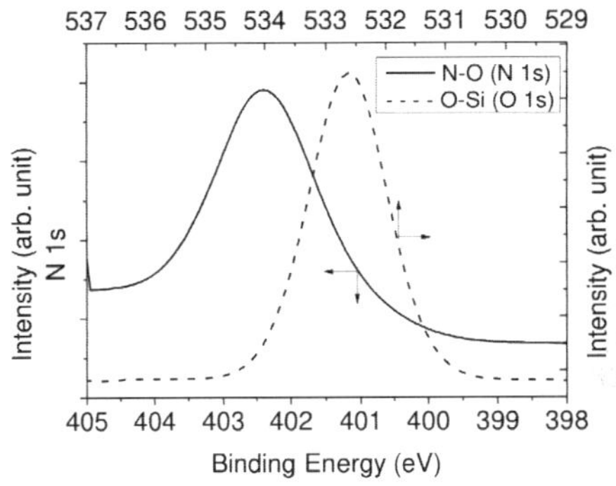

Figure 4. O 1s XPS spectra for the sample oxynitrided at 850°C during 40s.

XPS analysis allows one to obtain the atomic concentration at the surface region (around 4 nm in depth) for samples with thin layers besides to give the local chemical structure, from the signals associated to the chemical elements in these samples [8-10]. Figure 4 shows the XPS spectra where it is indicated the signals for N 1s and O 1s in expanded horizontal scale. The O 1s signal is mainly due to oxygen bonded to silicon (O-Si) at 532.6 eV, and the N 1s signal is compounded of nitrogen bonded to oxygen (N-O) at 402.4 eV, both pointing to the formation of SiO_xN_y grown by RTP. Based on the spectra exhibited in figure 4, the atomic concentration of nitrogen is 0.6% and the Si/O ratio in the oxynitride is approximately 1.9, which means an almost stoichiometric SiO_2 with a small amount of nitrogen.

Figures 5a and 5b shows, respectively, the RBS spectra for the SiO_2 reference and for the oxynitride grown at 850°C during 40s, both at the oxygen resonance-energy of 3.039 MeV. Using the areas under the oxygen peaks for the oxynitride (A_O) and for the 20nm-thick SiO_2 reference (A_R), the total planar concentration of the oxynitride was calculated from equation (1) and resulted 5.5 x $10^{15}cm^2$.

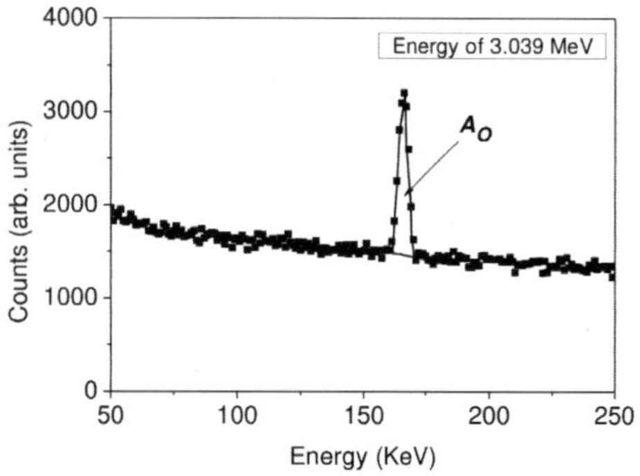

Figure 5a. RBS spectrum of the SiO_2 reference showing the oxygen resonant peak.

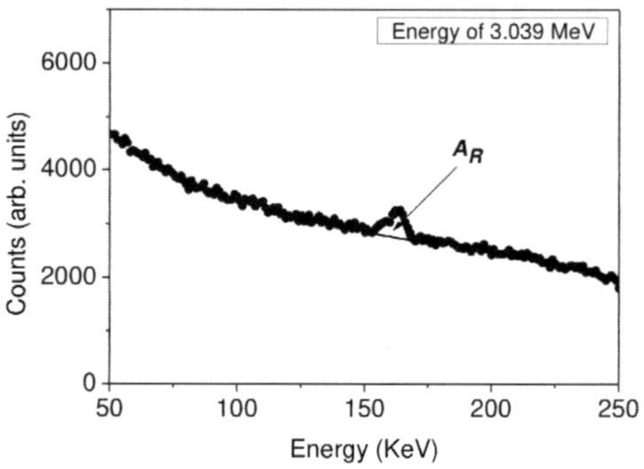

Figure 5b. RBS spectrum showing the oxygen resonant peak for the oxynitride grown at 850°C during 40s.

4. Conclusions

Thin silicon oxynitrides have been grown in a homemade RTP using a low mass quartz carrier. The thicknesses were in the range of 0.97 to 2.39nm with uniformity better than 0.4% in large areas of 3 inches silicon wafers. The samples processed at 700°C exhibited a rapid initial-oxidation starting from the processing time of 10 s with a growth rate of 0.12 nm/s and becomes approximately constant and equal to 0.0025 nm/s in the range of 40 to 160 s. This situation was modeled by an oxynitridation process dominated by surface reaction. For samples processed at 850°C, an approximately linear growth rate of 0.08 nm/s was observed in the range of 10 to 80s. XPS presented O 1s signal mainly due to oxygen bonded to silicon (O-Si) at 532.6 eV, and the N 1s signal of nitrogen bonded to oxygen (N-O) at 402.4 eV, both pointing to the formation of SiO_xN_y grown by RTP. On the other hand, the atomic concentration of nitrogen is 0.6 at% and the Si/O ratio in the oxynitride is approximately 1.9, which means an almost stoichiometric SiO_2 with a small amount of nitrogen. In addition, the planar concentration of oxygen was extracted from $^{16}O(\alpha, \alpha)$ elastic-scattering signal at 3.039MeV for the oxynitride grown at 850°C during 40s and resulted 5.5 x $10^{15}cm^2$.

5. Acknowledgements

The authors would like to thank CNPq for the financial support and Laboratório de Espectroscopia de Fotoelétrons from UNESP-Araraquara for the XPS measurements.

References

[1] Depas M, van Meirhaeghe R L, Laflère W H and Cardon F 1993 *Microelect. Eng.* **22** 61
[2] Chang K-M, Yang W-C, Chen C-F and Hang B-F 2004 *J. Electroch. Soc.* **151** F118
[3] Lu Z H, Tay S P, Cao R and Pianetta P 1995 *Appl. Phys. Lett*, **67** 2836
[4] Xu Y et al 2014 J. *Appl. Phys.* **115** 033502
[5] Kern W and Puotinen D A 1970 *RCA Rev.* **31** 187
[6] Scofield J H 1976 *J. Electron Spectrosc. and Related Phenomena* **8** 129
[7] Mezey G 1976 *Ion Beam Surface Layer Analysis* vol 1, ed O Meyer (New York: Plenum Press)
[8] Deal B E and Grove A S 1965 *J. of Appl. Phys.* **36** 3770–78
[9] Wagner C D, Riggs W M, Davis L E, Moulder J F and Mullenberg G E 1979 *Handbook of x-ray photoelectron spectroscopy* (Minnesota: Perking-Elmer Corporation)
[10] Kawase K, Tanimura J, Kurokawa H, Kobayashi K, Teramoto A, Ogata T and Inoue M 1999 *Materials Science in Semiconductor Processing* **2** 225
[11] Green M L, Gusev E P, Degraeve R and Garfunkel E L 2001 *J. of Appl. Phys. Reviews* **90** 2057

IOP Conf. Series: Materials Science and Engineering **76** (2015) 012003 doi:10.1088/1757-899X/76/1/012003

Nano-photocatalysts based on ZnS quantum dots/chitosan for the photodegradation of dye pollutants

H S Mansur[1] **and A A P Mansur**

Center of Nanoscience, Nanotechnology and Innovation-CeNano[2]I, Department of Metallurgical and Materials Engineering, Federal University of Minas Gerais, Av. Antônio Carlos, 6627, School of Engineering, 31270-901, Belo Horizonte/MG, Brazil.

E-mail: hmansur@demet.ufmg.br

Abstract. In this work, nano-photocatalysts based on ZnS quantum dots (QD) functionalized by chitosan were developed using "green" colloidal chemical process in aqueous media at room temperature. The ZnS/chitosan nano-photocatalysts were extensively characterized and the results demonstrated that chitosan was an effective capping ligand for the direct production of water-soluble ZnS QDs with average nanocrystal sizes of approximately 3.5 nm. Methylene blue dye was used as "model organic pollutant", which was effectively oxidized by the photocatalytic surface activity of the ZnS/chitosan nanostructured systems under UV irradiation. In summary, innovative "green" nano-photocatalyst nanomaterials were produced based on a fluorescent inorganic "core" of ZnS QDs and a biocompatible organic "shell" of chitosan for potential use on the photodegradation of hazardous dye pollutants present in industrial wastewater.

1. Introduction

Although international environmental standards are becoming more stringent, many industries, such as textile, plastics, paper and pulp, leather, cosmetics, and food, generate streams of waste effluents that contain high amounts of organic dyes, which exhibit carcinogenic and mutagenic properties towards living organisms and pose a serious threat to human life [1-3]. Despite several technological systems for the removal of organic pollutants have been developed, most of them are based on conventional processes for treatment of these effluents, which are unsatisfactory for purifying the wastewater from the industries. Therefore, traditional water treatment methods are now gradually being complemented by advanced oxidation processes (AOPs), such as photocatalysis methods that can provide quick and high efficiency on the oxidation of organic pollutants at low levels without the formation of polycyclic products [3,5]. Among the alternatives for "low-dimensional" materials for photocatalysis application, semiconductor nanocrystals (referred to as quantum dots, QDs) have recently emerged as a novel class of nanomaterials with unique physicochemical, electronic, magnetic and optical properties. Currently, the most commonly used QDs are made of Cd^{2+} core, which is highly toxic to humans and to the environment. Even though Cd^{2+} is incorporated into the nanocrystalline core as sulfides or selenides, surrounded by a "biocompatible shell" (e.g. polymers, proteins, polysaccharides), there is no definitive

[1] To whom any correspondence should be addressed.

information regarding how these toxic ions will impact the use of QDs in environmental and biomedical applications. Thus, "cadmium-free" nanomaterials, such as zinc sulfides (ZnS), have been interesting choices for producing QDs eco-friendly photocatalyst, due to the higher surface area-to-volume ratio as compared to their bulk counterparts with superior photon absorption at the nano-interfaces. In addition, size confinement caused by the small dimension of the QDs leads to a significant increase in the band gap energy compared to the bulk material producing a higher redox potential in the system [6].

In this paper, we endeavor to develop a novel nano-photocatalyst based on semiconductor ZnS QDs conjugated with chitosan by using a single-step "green chemistry" aqueous colloidal process at room temperature. In addition, methylene blue (MB) dye was used as "model organic pollutant", which was effectively oxidized by the photocatalytic surface activity of the ZnS/chitosan conjugates under UV irradiation. This new class of nanohybrids may offer countless opportunities to develop fluorescent and photocatalytic nanomaterials for environmental applications.

2. Experimental procedure

ZnS nanoparticles (QD-ZnS) were synthesized *via* an aqueous route at room temperature (RT) as previously reported by our group [7].

UV-Vis spectroscopy measurements were conducted using a Perkin-Elmer spectrometer (Lambda EZ-210) in transmission mode over a wavelength range of 600 nm to 190 nm. The absorption spectra were used to monitor the formation of ZnS QDs during the reaction as well as their relative colloidal stability in the medium.

Photoluminescence (PL) characterization of the QD-ZnS conjugate was conducted based on spectra acquired at RT using a Nanodrop 3300 fluoro-spectrometer (Thermo Scientific, UV Light Emitting Diode (LED) with $\lambda_{excitation}$= 365±10 nm).

QD-ZnS conjugate and chitosan were analyzed using diffuse reflectance infrared Fourier transform spectroscopy (DRIFTS) (Thermo Fischer, Nicolet 6700) over a range of 400 to 4000 cm^{-1} with 64 scans at a resolution of 2 cm^{-1}. These samples were prepared by placing a droplet of the chitosan solution or QDs colloidal dispersions onto the KBr powder followed by drying at 60±2 °C for 24 h.

Morphological and structural features of the ZnS QDs were characterized using transmission electron microscopy (G2-20-FEI transmission electron microscope, TEM) coupled to an energy dispersive X-ray (EDX) microprobe and selected area electron diffraction (SAED) analysis at an accelerating voltage of 200 kV. In all of the TEM analyses, the samples were prepared by dropping the colloidal dispersion onto a holey carbon/cupper grid.

The photocatalytic activity of the ZnS conjugates was evaluated *via* the photodegradation of MB (water-soluble dye that was used "model organic pollutant") under UV light irradiation (UVC, λ=280-100 nm, 6 W, at λ=254 nm, Boitton Instruments, parallel to the surface of the suspension at a fixed distance of 20 cm). The experiments were performed at RT using 15 mL of a suspension containing 30 $mg.L^{-1}$ of the QD-ZnS catalyst and 2.5 $mg.L^{-1}$ (QD-ZnS/dye = 12) of the dye which was poured into a petri dish (10 cm in diameter). The mixed suspensions were first magnetically stirred in the dark for 1.5 h to achieve adsorption-desorption equilibrium. At certain time intervals, 3 mL of the mixed suspensions were collected and analyzed by UV-Vis spectroscopy (Lambda EZ-210, Perkin Elmer) to determine the dye concentration based on the Beer-Lambert curve to correlate the absorbance with the dye concentration at its characteristic maximum absorption wavelength (λ=664 nm).

3. Results and Discussion

In figure 1(a), it is presented the UV-vis spectrum of ZnS semiconductor (ZnS QDs) conjugates using the carbohydrate chitosan as the stabilizing ligand during the synthesis. The spectrum exhibits a broad absorption band between 250 and 350 nm with the exciton absorption transition (λ_{exc}) at approximately 267 nm indicating that the ZnS QDs were produced and stabilized by chitosan as the capping-ligand within the quantum confinement regime.

The optical band gap (absorbance onset), the blue-shift values and QD size [8] were determined from the absorption coefficient data as a function of the wavelength using the Tauc relation (figure 1(b)) extracted from the UV-Vis spectra [9]. As summarized in figure 2, the results support the formation of ZnS-chitosan conjugates in aqueous media because the band gap energy of the semiconductor (3.92 eV, that corresponds to 316 nm) was significantly greater than the ZnS bulk (i.e., 3.61 eV, blue-shift = 0.42 eV) [10].

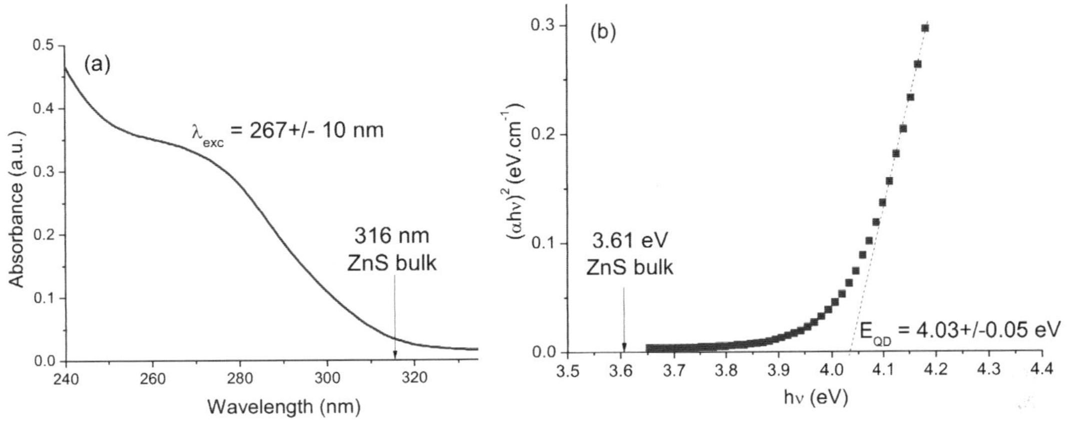

Figure 1. (a) UV-Vis spectra and (b) optical band gap plot using Tauc equation of the ZnS-chitosan conjugates.

Figure 3A shows the photoluminescence spectrum collected at RT of the system under evaluation. The band edge recombination in UV range was not detected and other bands in violet-blue range were observed. According to the literature, ZnS quantum dots typically exhibit emission peaks in the 400-550 nm wavelength range, which is primarily associated with point defects, such as vacancies (V) and interstitial ions (I) and also surface defects [7].

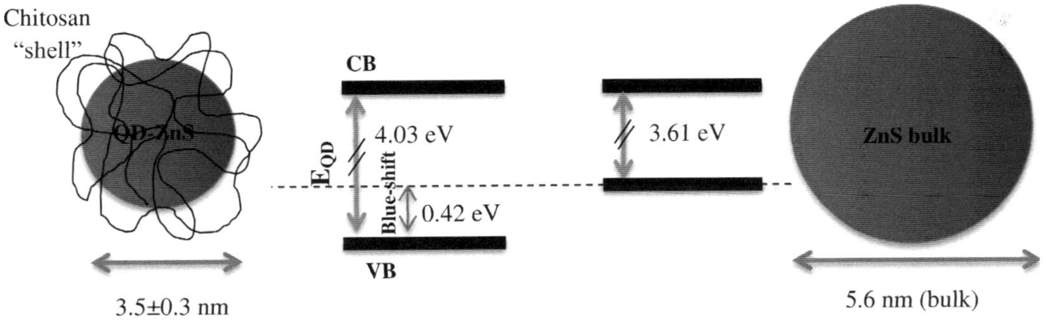

Figure 2. Schematically representation of size and band gap energy of ZnS QDs in comparison to ZnS bulk.

FTIR spectrum of chitosan (figure 3B(a)) presents absorption peaks at 1645 cm^{-1} and 1560 cm^{-1}, which are assigned to the carbonyl stretching of the secondary amides (amide I band) and the N-H bending vibrations of the deacetylated primary amine (-NH_2) and amide II band, respectively. In addition, the absorptions at 1030-1040 cm^{-1} and 1080-1100 cm^{-1} indicate the C–O stretching vibration in chitosan, which are associated with the C6-OH primary alcohol and the C3-OH secondary alcohol respectively [7]. After conjugating the ZnS QDs dots with the capping chitosan (figure 3B(b)) all these bands of chitosan were "red-shifted", i.e., shifted to lower wavenumber. These changes can be mainly attributed to the interactions occurring between the functional groups of the chitosan ligand and the ZnS quantum dots, indicating that the primary and secondary alcohols, amine, and acetamide groups of the chitosan polymer have interacted with the ZnS nanocrystals at the surfaces [10].

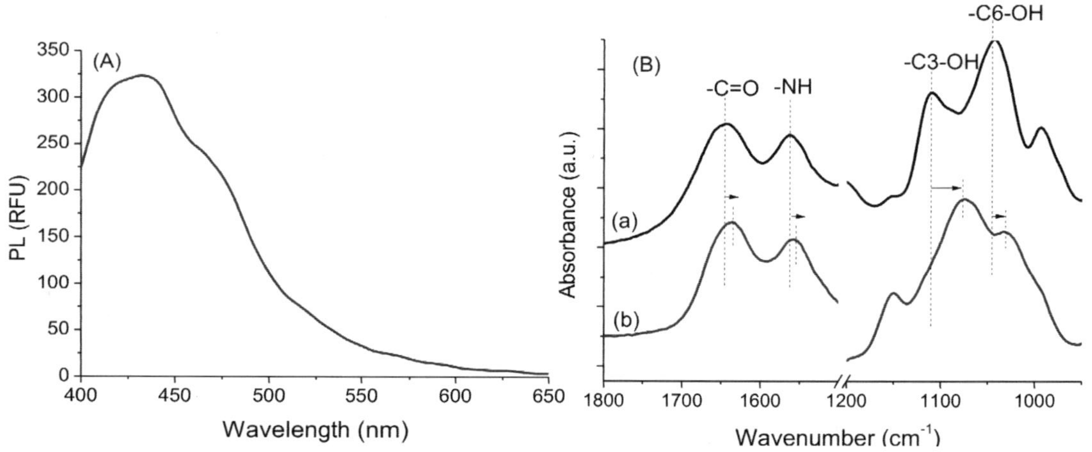

Figure 3. (A) PL spectrum of ZnS-chitosan conjugates. (B) FTIR spectra of chitosan (a) and QD-ZnS conjugate (b).

Figure 4 shows representative TEM images of ZnS QDs with spherical-like shapes. The estimated nanoparticle size is in reasonable agreement with the values obtained from the UV-Vis optical absorbance (3.5±0.3 nm). In addition, the electron diffraction pattern of the QDs with a lattice parameter compatible with the ZnS cubic crystal (JCPDS 05-0566) is showed in figure 4(A). EDX spectrum shows the chemical analysis of the nanocrystals with Zn and S as the major elements (figure 4(B)), excluding the copper, oxygen and carbon peaks related to the TEM grid and the polymer stabilizer.

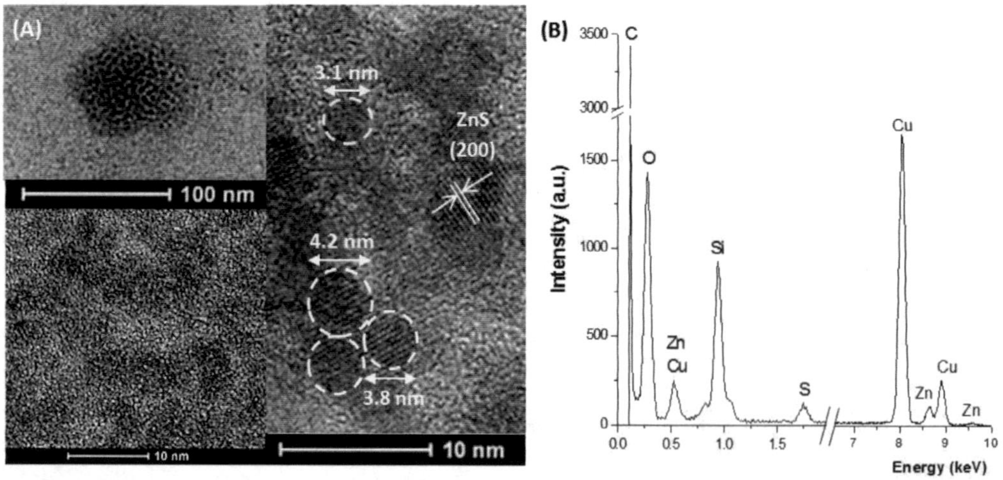

Figure 4. (A) TEM image with nanocrystal plane spacing and (B) EDX spectrum of QD-ZnS.

The ZnS-chitosan conjugates were tested as catalysts in the presence UV radiation and the results are presented in figure 5. The discoloration profile at 664 nm (figure 5(a)) for the solution containing methylene blue dye indicated that the ZnS QDs removed the dye from the aqueous medium effectively: approximately 90% of the coloration after a 90 min reaction under UV light (figure 5(b). This photocatalytic performance of the ZnS QDs conjugates was expected by considering: (i) ZnS QD band gap energy and relative position; (ii) the redox potential of $H_2O/OH^{\bullet}$, $OH^{-}/OH^{\bullet}$ and $O_2/O_2^{\bullet -}$ species that lies within the band gap of the ZnS QDs; and (iii) the intermediate energy states due to defects in crystal structure inhibiting or delaying the rapid carrier recombination processes [4,11].

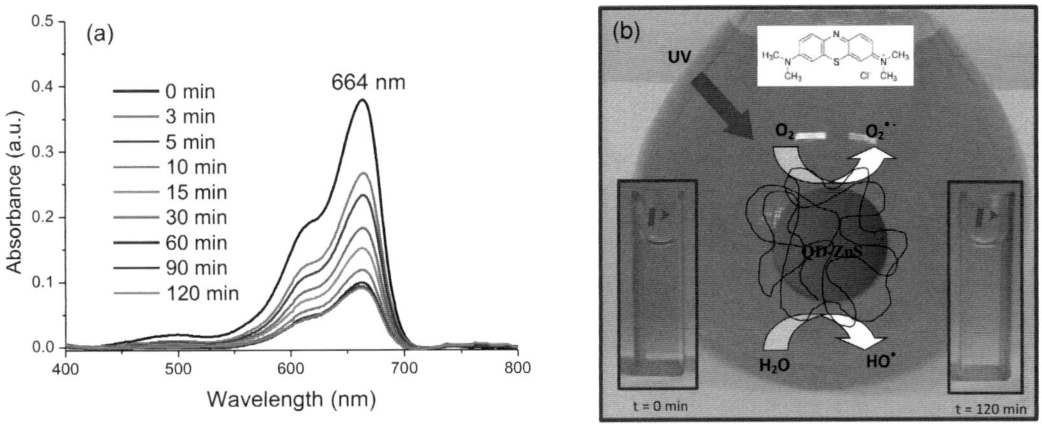

Figure 5. (a) Absorption spectral changes of MB aqueous solution in the presence of ZnS QDs under UV light (ZnS QDs:MB = 12). (b) Schematic MB degradation by ZnS-chitosan nano-photocatalyst.

4. Conclusions

Nanoconjugates based on ZnS quantum dots stabilized by biocompatible chitosan ligands were effectively synthesized *via* a novel facile route based entirely on aqueous colloidal chemistry at room temperature. The ZnSichitosan conjugates were fluorescent and exhibited photocatalytic activity for the degradation of organic pollutant compounds dyes (i.e., methylene blue) under UV irradiation.

Acknowledgments

The authors acknowledge the financial support from Brazilian agencies CAPES, FAPEMIG and CNPq. The authors are grateful to the staff of the Microscopy Center/UFMG for the TEM analyses. .

References

[1] Vinu R and Madras G 2010 *J. Indian Inst. Sci.* **90** 189
[2] Lachheb H, Puzenat E, Houas A, Ksibi M, Elaloui E, Guillard C and Herrmann J –M 2002 *Appl. Catal. B: Environ.* **39** 75
[3] Rauf M A, Meetani M A, Khaleel A and Ahmed A 2010 *Chem. Eng. J.* **157** 373
[4] Mansur A A P, Mansur H S, Ramanery F P, Oliveira L C A and Souza P 2014 *Appl. Catal. B: Environ.* **158-159** 269
[5] Rajabi H R, Khani O, Shamsipur M and Vatanpour V 2013 *J. Hazard. Mater.* **250-251** 370
[6] Qu X, Alvarez P J J and Li Q 2013 *Wat. Res.* **47** 3931
[7] Ramanery F P, Mansur A A P and Mansur H S 2013 *Nanoscale Res. Lett.* **8** 512
[8] Jaiswal A, Sanpui P, Chattopadhyay A and Ghosh S S 2011 *Plasmonics* **6** 125
[9] Tauc J and Menth A 1972 *J. Non-Cryst. Solids* **8/10** 569
[10] Cooper J K, Franco A M, Gul S, Corrado C and Zhang J Z 2011 *Langmuir* **27** 8486
[11] Oliveira L C A, Oliveira H S, Mayrink G, Mansur H S, Mansur A A P and Moreira R L 2014 *Appl. Catal. B: Environ.* **152-153** 403

Vulcanization reaction of squalene and S_8 powder studied by Sulfur K-edge NEXAFS under liquid phase

S Yagi[1,2,3], Y Menjo[2], C Tsukada[2], S Ogawa[2], G Kutluk[3], H Namatame[3] and M Taniguchi[3]

[1]Green conversion section, EcoTopia Science Institute, Nagoya University, Nagoya 464-8603, JAPAN
[2] Graduate School of Engineering, Nagoya University, Nagoya 464-8603, JAPAN
[3] Synchrotron radiation Center, Hiroshima University, Higashi-Hiroshima 739-0046, JAPAN

E-mail: s-yagi@nucl.nagoya-u.ac.jp

Abstract. Vulcanized rubber materials are useful in our surroundings. However, detail structure and reaction are not revealed even in present. Since squalene molecule possesses some same properties compared with natural rubber, we have prepared the samples of vulcanized squalene at 140 °C for several hours. To understand the vulcanization reaction,sulfur K-edge NEXAFS measurements have been carried out for the vulcanized squalene under liquid phase with He-path system and fluorescence detection mode. Moreover, we have tried curve fitting analysis of NEXAFS spectra. The results indicate that the squalene has been vulcanized by the S_8 molecule at 140 °C and the S_8 molecule length is shortened from 8 to 5-6 after the vulcanization reaction.

1. Introduction

Vulcanization reaction has been invented by C. Goodyear in 1839 [1]. Natural rubber is vulcanized by sulfur powder and the vulcanized rubber materials have been used historically in our surroundings. The rubber materials possess unique physical properties such as high elasticity and restoring force, and can be used to the bio-functional materials. Furthermore, the rubber materials are also considered to apply to bio-implant because it is chemically stable. For example, the vulcanized rubber, which possesses some interactions with the solid state surface, can be used in many industrial fields. Therefore, the vulcanized rubber has many possibilities and is paid attention in recent years.

Though the natural rubber consists of polymers of the organic compound isoprene (cis-C_5H_8), the natural rubber includes minor impurities of other organic compounds as protein and fatty acid molecules. It is thought that the properties of rubber materials are influenced greatly by the existence of both the structural difference and the impurities. Therefore, the control of the properties of the rubber materials is very difficult [2,3,4].

The rubber materials are produced by heating the mixture of raw rubber classified in terms of high polymerized hydrocarbon compound, sulfur powder (S_8) and other added materials. The S_8 reacts with the carbon chains in the raw rubber. This reaction is called "bridged reaction". The physical properties of the rubber materials depend upon the three-dimensional structure by the bridged reaction. Generally, the reaction and the structure are known as "vulcanization reaction" and "bridge structure", as shown in figure 1, respectively. Even in present time, the mechanism of the vulcanization reaction has not been firmly resolved. In that sense, we are thinking the vulcanization reaction is complex. To understand the vulcanization reaction and the bridge structure is important to synthesize the vulcanized rubber materials. However, it is very difficult to mix the vulcanization reaction reagent with the raw rubber in the laboratory by its high viscosity.

Squalene has some specific properties, which are liquid phase hydrocarbon compound, same monomer of natural rubber and less impurities. Take into account of those factors, we propose the squalene instead of the raw rubber in this study. But there is a few manuscripts written about the vulcanization reaction with the squalene [5,6]. The purpose of this study is to understand the vulcanization reaction for the bridge structure with reaction time variation using sulfur K-edge NEXAFS (near-edge X-ray absorption fine structure) measurement.

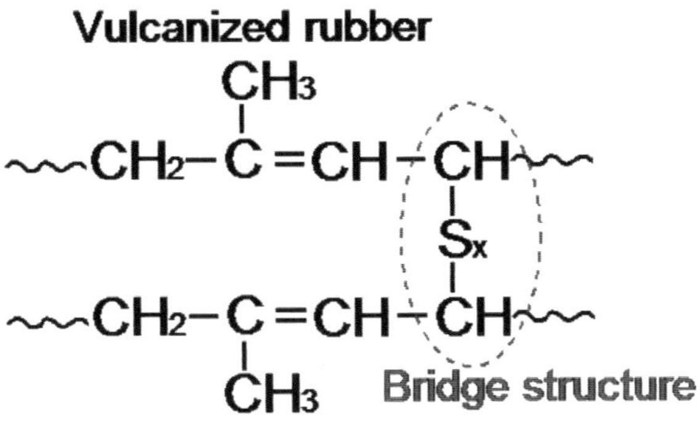

Figure 1. Schematic view of vulcanized rubber and bridge structure by S_x (x=1-8).

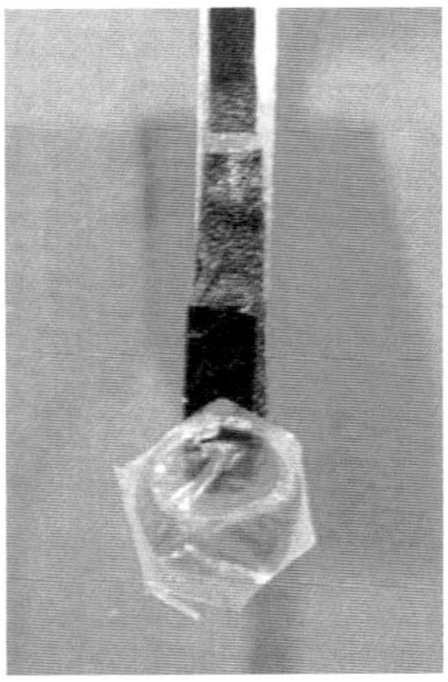

Figure 2. Photographic view of vulcanized squalene with polyethylene film cell for NEXAFS measurement.

2. Experimental

The reagents of the squalene (2.5 g: KATAYAMA CHEMICAL INDUSTRIES Co., LTD.) and the S_8 powder (50 mg: KATAYAMA CHEMICAL INDUSTRIES Co., LTD.) were mixed into closed vials under Ar atmosphere to avoid the air oxidation. The mixed samples were stirred at 140 °C for several hours (0, 3, 6, 9 and 12 h). Sulfur K-edge NEXAFS measurement of the vulcanized squalene samples was carried out by fluorescence X-ray yield mode using the atmospheric XAFS measurement system with He-path at the BL-3 on HiSOR [7] and the BL6N1 on Aichi Synchrotron Radiation Center (AichiSR) beamlines. Those soft X-ray beamlines were used the InSb(111) crystals in the monochromator system. Especially, the synchrotron facility of the electron storage ring in AichiSR was operated at the top-up mode. The photon energy was calibrated on the assumption that the first peak of K_2SO_4 powder appears at 2481.70 eV. The fluorescence yield detections at BL-3 and BL6N1 were employed using a UHV-compatible gas-flow type proportional counter with P-10 gas (10 % CH4 in Ar) and a SDD (silicon drift detector), respectively. We used polyethylene film cell with about 12 μm thickness for liquid samples, as shown in figure 2. In fact, a slight sulfate contaminant such as SO_3 and/or SO_4 was observed in the film cell. However, because the sulfur component amount contained in the measurement sample was large, a large problem did not occur. The sulfur K-edge NEXAFS spectrum for the contaminants of SO_3 and SO_4 of the film cell were shown in figure 3.

3. Results and disscussion

Figure 3 shows sulfur K-edge NEXAFS spectra for the vulcanized squalene samples reacted at 140 °C for several hours and the background of contaminants described in Experimental section. All NEXAFS spectra, except BG, are normalized by the edge-jump value. Moreover, an inset is also shown by magnified photon energy area. The structures of BG spectrum mean that some sulfur contaminants as thiol and sulfate are existing in the cell membrane, but the amount of those contaminants are extremely little. Thus we conclude that the BG structures never affect the NEXAFS spectra of vulcanized squalene. In our study, though we have detected the sulfur signals by use of the fluorescence detector, usually this detection method produces a self-absorption effect [8,9,10]. According to the self-absorption effect, the peak intensity is including some error. The first peak is observed at 2471.8 eV in start phase, which means the reaction time for vulcanization is 0 hour in figure 3. The peak intensity is approximately 4.4 compared with the edge-jump value. This value is almost same as the result reported by I.J. Pickering *et al.* and G.N. George *et al.* [11,12]. This comparison indicates that our NEXAFS spectra include little the self-absorption effect in spite of using the fluorescence detection. Thus, the S_8 materials are existing under highly dispersed condition in the squalene. In other words, when someone obtains XAFS spectra of highly dispersed molecularly or atomically by means of the fluorescence detection, there is no self-absorption effect.

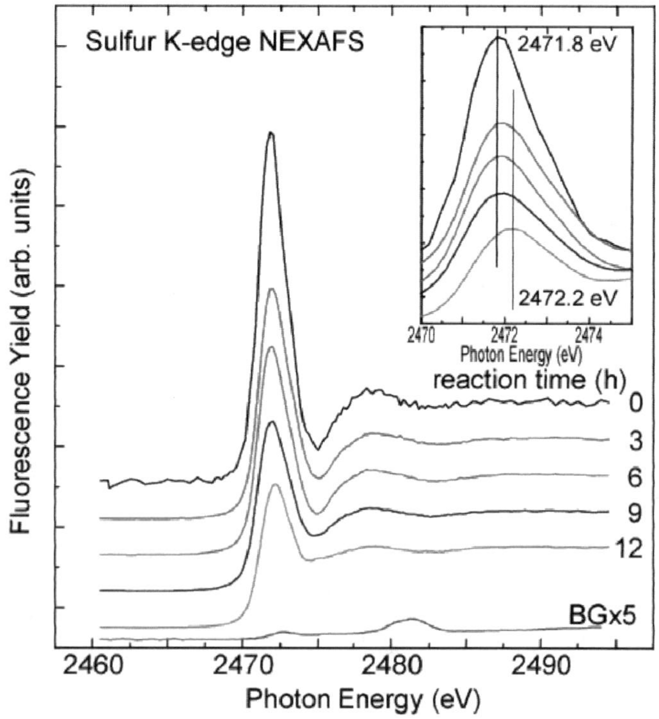

Figure 3. Sulfur K-edge NEXAFS spectra for vulcanized squalene at 140 °C for 0-12 hours and background (BG) of film cell. Inset shows magnified spectra of first peak area.

The first peak at 2471.8 eV is assigned to the transition from sulfur 1s to anti-bonding molecular orbital σ*(S-S). But this peak position is not same as the position of S_8 powder of 2471.6 eV, which is not shown in figure 3. It is thought that the bond length of S-S in S_8 powder in squalene matrix is slightly shorter than that of S_8 powder in vacuum. The first peak shifts to 2472.2 eV continuously with reaction time in 12 hours. In addition, the peak position after 12 hours does not match the position of the monosulfide vulcanized squalene of σ*(S-C) at 2472.8 eV reported by S. Inoue *et al.* [13]. This result means that the chemical state of vulcanized squalene after 12 hours is not single chemical state, but the mixed materials of S_8 and monosulfide states. Considering that the intensity of the peak is decreased gradually proportional to the reaction time, the spectrum change means the specific reaction. Concretely, the reduce of first peak intensity and the shift of peak position suggest the decrease of the number of S-S bond and the creation of S-C bond.

It is necessary to carry out the curve fitting of the NEXAFS spectra to estimate quantitatively the decrease and the increase the peak intensities correspond to the S-S or S-C bond element. The curve fitting needs the energy position of ionization potential of sulfur 1s level. Usually the ionization potential is measured by means of XPS (X-ray photoelectron spectroscopy) technique. But we cannot measure the chemical state of the sulfur, since our samples are liquid phase. Thus, we have performed the curve fitting about the simple XAFS spectrum of the S_8 powder dispersed into squalene. The result of the curve fitting for the S_8 powder dispersed into squalene at 120 °C, not shown here, the ionization potential position is estimated to be 2476.0 eV. Therefore, we have tried to the curve fitting of all NEXAFS spectra by use of the energy value of 2476.0 eV as a step function. Since the S-S and S-C bonds in the vulcanization reaction exist under liquid environment, we have adopted an asymmetric Gaussian peak in curve fitting analyses.

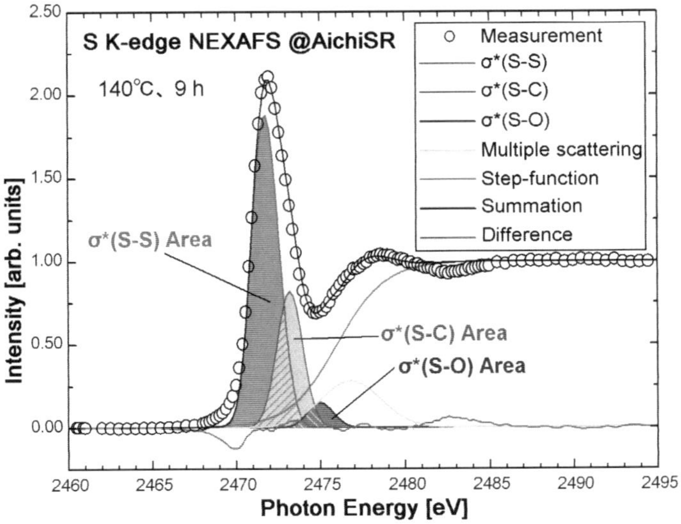

Figure 4. Sulfur K-edge NEXAFS spectrum for vulcanized squalene at 140 °C for 9 hours, and also showing fitting result.

Figure 4 shows a typical sulfur K-edge NEXAFS spectrum, which is of the vulcanized squalene at 140 °C for 9 hours, and the result of curve fitting. Three peak components, which are assigned as σ*(S-S), σ*(S-C) and σ*(S-O) resonances from lower photon energy side, have been extracted. Two peaks of σ*(S-S) and σ*(S-C) are originated from the S8 source and the vulcanized bridge structure, respectively. In addition, unfortunately we have observed the σ*(S-O) peak, which are occurred by the oxidation reaction with some contaminants of water in squalene and/or air.

We have performed the curve fitting of NEXAFS spectra for all vulcanized squalene at 140 °C and extracted the intensities of three peak components related to S-S, S-C and S-O bonds. Those results are shown in figure 5. We have carried out the normalization which sum of the three peak components is to be 1.0. Judging from the results of curve fitting, the peak intensity of S-S bond decreases gradually in proportional to the reaction time. On the other hand, the S-C peak intensity increases. These results mean that the vulcanization reaction is occurring at the carbon of squalene. When the fitting line related with the S-S bond is extrapolated, the peak intensity approaches 0.6. It seems that this value means the 8 value of S_8 becomes 5 (S_5) or 6 (S_6).

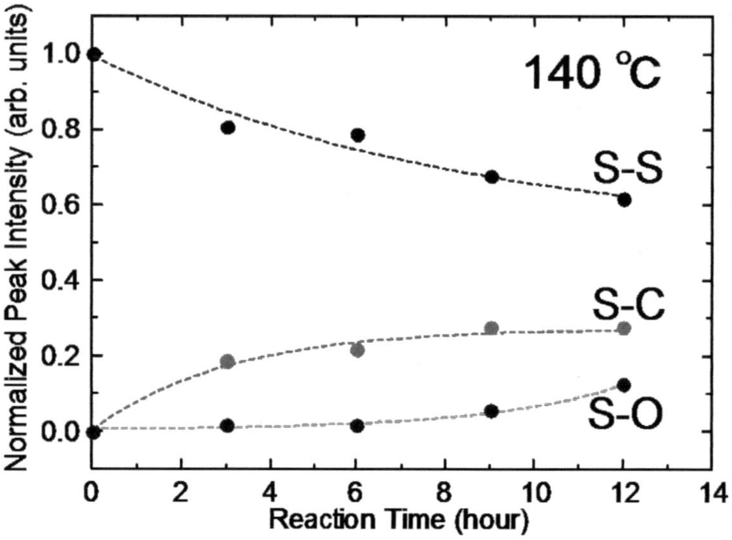

Figure 5. Curve fitting results of Sulfur K-edge NEXAFS spectra for vulcanized squalene at 140 °C.

4. Conclusion

We have prepared the samples of vulcanized squalene at 140 °C for several hours instead of the vulcanized rubber materials. The sulfur K-edge NEXAFS spectra for the samples have been obtained under liquid phase with He-path system and fluorescence detection mode. All NEXAFS spectra have not shown very little the self-absorption effect. Judging from the results of curve fitting of NEXAFS spectra, the squalene has been vulcanized by the S_8 molecule at 140 °C. In addition, the S_8 molecule length is shortened from 8 to 5-6 in proportional to the reaction time. It can be said that the sulfur K-edge NEXAFS analysis using He-path system is very powerful tool to understand the bridge structure of vulcanization reaction.

Acknowledgement

We would like to thank Dr. Nomoto T for valuable supports at BL6N1 in AichiSR. This work was performed under the approvals of AichiSR (Nos. 2503023 and 2503026) and HSRC Program Advisory Committee (Nos. 13-A-26 and 14-A-54). The authors are grateful for the financial support of a Grant-in-Aid for Scientific Research Nos. 253655 and 26289255 from the Ministry of Education, Science, Sports and Culture, Japan.

References

[1] Goodyear C 1853 *Gum-elastic and Its Varieties: With a Detailed Account of Its Applications and Uses, and of the Discovery of Vulcanization* (New Haven) p22
[2] Zaper A M and Koeing J L 1988 *Makromol Chem.* **189** 1239
[3] Mukhopadhyay R, De S K and Chakraborty S N 1977 *Polymer* **18** 1243
[4] Suzuki T, Osaka N, Endo H, Shibayama M, Ikeda Y, Asai H, Higashitani N, Kokubo Y and Kohjiya S 2010 *Macromolecules* **43** 1556
[5] Taweepreda W, Nu-Mard R, Pattanasiriwisawa W and Songsiriritthigul P 2009 *Journal of Physics: Conference Series* **190** 012150
[6] Gros M, Borrós S, Amabilino D B, Veciana J and Folch I 2001 *J. Mass Spectrometry* **36** 294
[7] Yagi S, Kutluk G, Matsui T, Matano A, Hiraya A, Hashimoto E and Taniguchi M 2001 *Nuclear Instruments and Methods in Physics Research* A **467** 723
[8] Tröger L, Arvanitis D, Baberschke K, Michaelis H, Grimm U and Zschech E 1992 *Physical Review* B **46** 3283

[9] Pfalzer P, Urbach J -P, Klemm M, Horn S, denBoer M L, Frenkel A I and Kirkland J P 1999 *Physical Review* B **60** 9335

[10] Klysubun W, Sombunchoo P, Wongprachanukul N, Tarawarakarn P, Klinkhieo S, Chaiprapa J and Songsiriritthigul P 2007 *Nuclear Instruments and Methods in Physics Research* A **582** 87

[11] Pickering I J, George G N, Yu E Y, Brune D C, Tuschak C, Overmann J, Beatty J T and Prince R C 2001 *Biochemistry* **40** 8138

[12] George G N, Pickering I J, Yu E Y and Prince R C 2002 *Microbiology* **148** 2267

[13] Inoue S 1983 *Nippon Gomu Kyokaishi* **56** 686

Bio-functionalized water-soluble ZnS quantum dots using carboxymethylchitosan

A A P Mansur, H S Mansur[1], F G L M Borsagli and F P Ramanery

Center of Nanoscience, Nanotechnology and Innovation-CeNano[2]I, Department of Metallurgical and Materials Engineering, Federal University of Minas Gerais, Av. Antônio Carlos, 6627, School of Engineering, 31270-901, Belo Horizonte/MG, Brazil.

E-mail: hmansur@demet.ufmg.br

Abstract. The major goal of this study was to develop an innovative green route for synthesizing biocompatible water-soluble luminescent QDs using chemically modified chitosan as the ligand in aqueous media. The preparation of ZnS QDs bio-functionalized by carboxymethylchitosan (CMC) was performed using a single-step aqueous colloidal process at room temperature. The results showed that water-dispersible ZnS nanocrystals capped by CMC were produced within the quantum-size confinement regime. Moreover, the luminescent properties of ZnS QDs were significantly affected by the pH during the synthesis due to the size distribution of the nanoparticles and their density of surface states.

1. Introduction

Recent advances in nanotechnology and nanoscience for biological and medical applications have been the major driving force in the creation of new nanohybrids and nanoconjugates. Essentially, to be considered as nanosized, the systems must present one or more components with at least one dimension ranging from 1 to 100 nm. Besides the low dimensions, nanohybrids and nanoconjugates may bring together the intrinsic functionalities of inorganic nanoparticles combined with the biocompatible surfaces and interfaces offered by biomolecules and biopolymers, such as carbohydrates, proteins, and enzymes [1-2]. Among several alternatives of inorganic nanoparticles, semiconductor nanocrystals (or quantum dots, QDs) have emerged as an attractive nanomaterial owing to their unique physico-chemical, electronic, and optical properties. However, for potential applications of QDs in nanomedicine and biology their cytotoxicity must be eliminated because they are predominantly made of highly toxic heavy metal cores [2]. Presently, the most commonly used QDs contain divalent cadmium (Cd^{2+}), widely known as a toxin, due to its accumulation in living tissues and organs [3-4]. In that sense, "cadmium-free" nanomaterials are very promising possibilities, such as zinc divalent cations (Zn^{2+}), due to their natural environmental abundance and they are commonly found in several living organisms as crucial metabolic species. Another interesting approach has been used for producing QDs with biocompatible surfaces and interfaces based on capping ligands such as biomolecules and biopolymers rendering them biocompatibility for biomedical applications [3-5]. In view of the large number of available options of biocompatible

[1] To whom any correspondence should be addressed.

ligands to produce QD nanoconjugates for bio-applications, such as carbohydrates, proteins, and enzymes, chitosan and its derivatives have been often chosen, due to their biocompatibility, physico-chemical and mechanical properties, and relative chemical solubility and stability in aqueous medium [3-5]. Chitosan (poly-β(1→4)-2-amino-2-deoxy-D-glucose) is one of the most abundant polysaccharides (semi-processed) from natural sources, second only to cellulose [3-5] but it has very low solubility at physiological environment (pH~7.0). Consequently, chitosan derivatives such as carboxymethylated chitosans (O-carboxymethyl chitosan, O-CMC and N-carboxymethyl chitosan, N-CMC) may be good candidates as ligands for preparing biocompatible colloidal semiconductor nanoparticles because they are water-soluble in acidic, alkaline and physiologic media [5,6]. These combined features of chitosan and its derivatives offer limitless possibilities for developing water-dispersible QD conjugates for biological, pharmaceutical [2,5] and environmental purposes [7].
Hence, this research focused on the synthesis of ZnS quantum dots directly capped by carboxymethyl chitosan (CMC) using a green, facile, reproducible, and single-step aqueous process at room temperature.

2. Experimental

2.1. Materials
All reagents and precursors, $ZnCl_2$ (Sigma-Aldrich, USA, ≥ 98%), $Na_2S\cdot 9H_2O$ (Synth, Brazil, >98%), NaOH (Merck, USA, ≥ 99%), CH_3COOH (Synth, Brazil, ≥ 99.7%), HCl (Sigma-Aldrich, USA, 36.5-38.0%), monochloroacetic acid (Sigma-Aldrich, USA, 99%, $ClCH_2COOH$), ethanol (Synth, Brazil, 99.8%), and isopropanol (Sigma-Aldrich, USA, 99.5%) were used as-received. Chitosan powder (Aldrich, USA, MM = 310,000 to >375,000 g/mol, DD ≥ 75.0%, and viscosity 800-2000 cPoise, at 1 % in 1 % acetic acid) was used as the reference ligand. Deionized water (Millipore Simplicity™) with resistivity of 18 MΩ·cm was used in the preparation of all solutions and the syntheses were performed at room temperature (R.T. = 23±2 °C) unless specified.

2.2. Synthesis of carboxymethylchitosan (CMC) and ZnS quantum dots (QD_CMC and QD_CHI)
The method used to prepare carboxymethylchitosan at R.T. was based on the process previously reported by our group [2,4]. In brief, the chitosan (CHI) was chemically modified using the reaction with monochloroacetic acid in highly alkaline (NaOH) medium of isopropanol for approximately 4 h. In the sequence, the suspension was neutralized (HCl), filtered and the solid filtrate rinsed in ethanol to neutral, and then vacuum dried, leading to the formation of CMC flakes. Finally, the CMC solution (1.0% w/v) was prepared by dissolving 0.5 g of CMC flakes in 50 mL of deionized water under moderate magnetic stirring until complete solubilization occurred. ZnS nanoparticles were synthesized *via* an aqueous route at R.T. using CMC or CHI as the ligand, the zinc precursor solution ($ZnCl_2$, $8x10^{-3}$ $mol.L^{-1}$) and the sulfur source solution ($Na_2S\cdot 9H_2O$, $1.0x10^{-2}$ $mol.L^{-1}$) which were added to the reaction flask (S:Zn = 1:2) and stirred for 1 h. ZnS QDs colloidal suspensions were referred to as QD_CMC-pH4, QD_CMC-pH10 and QD_CHI-pH4 as a function of the pH and ligand of the QDs synthesis. Sampling aliquots were collected at different time intervals for Ultraviolet-Visible spectroscopy (UV-vis) measurements to evaluate the colloidal nucleation and stability.

2.3. Characterization of CMC and QD_CMC and QD_CHI colloidal quantum dots
The characteristic functional groups of chitosan, CMC and conjugates were evaluated by Fourier transform infrared spectroscopy (ATR-FTIR, in the range of 650 to 4000 cm^{-1}, Thermo Fischer, Nicolet 6700, 16 scans, 2 cm^{-1} resolution, with a background subtraction). The water solubility of CMC as a function of the pH was evaluated by the turbidity method using UV-vis in the range 190 nm $\leq \lambda \leq$ 800 nm based on the literature [8], which considered the sample insoluble when the transmittance was ≤ 85% (λ = 450 nm). The nucleation, growth and stabilization of colloidal ZnS QDs in CMC were monitored by UV-vis, and their physicochemical and optical properties were assessed using photoluminescent spectroscopy (PL). ZnS conjugates fluorescence emissions were measured at

R.T. using Nanodrop 3300 fluorospectrometer (Thermo Scientific, UV LED, $\lambda_{excitation}$= 365±10 nm). Nanostructural characterizations of the QD nanoconjugates were based on the images and selected area electron diffraction patterns (SAED) obtained using a Tecnai G2-20-FEI transmission electron microscope at an accelerating voltage of 200 kV. Energy-dispersive X-ray spectra (EDX) were collected using the TEM for element chemical analysis. In all of the TEM analyses, the samples were prepared by dropping the colloidal dispersion onto a porous carbon grid. The QD size and size-distribution data were obtained based on the TEM images by measuring at least 100 randomly selected nanoparticles using an image processing program (ImageJ, version 1.44, public domain, National Institutes of Health). All of the experiments were conducted in triplicates (n=3) unless otherwise noted. The data are presented as the mean±standard deviation.

3. Results and Discussion

The FTIR spectra of chitosan (CHI) and carboxymethylchitosan (CMC) are showed in figure 1(a) and figure 1(b), respectively. Unlike the CHI spectrum, CMC spectrum presented a characteristic band at approximately 1731 cm^{-1}, which is assigned to the carboxyl (–COOH) groups. The protonation of the amine group in CMC causes the appearance of two new bands, the first band at 1640-1620 cm^{-1}, associated with the antisymmetric stretching of $-NH_3^+$ overlapping the amide I band, and the second band at 1550-1500 cm^{-1} is related to the $-NH_3^+$ symmetric deformation, indicating the acid-form of CMC (H-CMC) [4,8].

The solubility tests based on UV-vis turbidity test indicated that CMC solutions were soluble under acidic and alkaline conditions, and only insoluble in small range of pH $6.0 < pH < 7.5$. That is one of the major advantages of CMC derivative compared to chitosan because it is soluble in both acid and alkaline aqueous media and chitosan is insoluble at pH higher than 6.5.

The UV-vis absorption spectra of the colloidal ZnS nanoparticles produced using CHI and CMC as the stabilizing ligand are showed in figure 2. In general, the curves exhibit a broad absorption band in the range from 250 to 350 nm, associated with the first excitonic transition indicating nucleation/growth of ZnS nanoparticles within the "quantum confinement regime". It was observed the energy "blue-shift" for the all the ZnS-conjugates curves as compared with the "bulk" value (figure 2, arrow at λ= 343 nm) [9]. Therefore, the optical band gap energy of ZnS QD (E_{QD}) is larger than that of the original bulk material (E_g). The optical band gap (absorbance onset) and the "blue-shift" values were determined from the absorption coefficient data as a function of wavelength using the "Tauc relation" [10] extracted from the UV-vis spectra. The results support the formation of ZnS QDs in the CHI and CMC media because the band gap energy values of QDs (E_{QD} equal to 3.74±0.02 for QD_CHI-pH4 and 3.83±0.02 eV and 4.10±0.02 eV for QD_CMC-pH4 and QD_CMC-pH10, respectively) were significantly greater than the ZnS bulk (i.e., 3.61 eV) [4]. The average ZnS nanoparticle sizes (2R) were estimated from the empirical model reported [4,11] using the optical band gap value (E_{QD}). The ZnS QDs were produced and stabilized by CMC with estimated diameters (2R) of 4.2±0.1 nm and 3.2±0.1 nm at pH 4.0 and 10.0, respectively, compared to the larger size 4.7±0.1 nm found for ZnS-CHI conjugate (at pH 4.0).

Figure 3 shows the photoluminescence spectra of the QD_CHI and QD_CMC colloidal nanoparticles collected at R.T. with their relative intensities. Based on the absorbance curves and the band gap energies evaluated under excitation, QD_CHI and QD_CMC nanoconjugates were expected to emit light in the UV range ($E_g \geq 3.6$ eV). However, band edge recombination was not detected, and other bands in the violet-blue range were observed (inset, figure 3). These results are in agreement with the reported emission spectra of ZnS nanoparticles, which typically exhibit radiative recombination in the 400-550 nm wavelength range arising from trap states associated with point and surface defects [4,12]. The emission band identified at approximately 416 nm is due to transitions involving Zn^{2+} interstitial states (I_{Zn}), while the emission band at 435 nm is assigned to sulfur vacancy states (V_S), compatible with the procedure used in this work using a stoichiometric molar ratio of $Zn^{2+}:S^{2-}$= 2:1. The band at approximately 470 nm, observed for all systems (independent to the pH of the synthesis), may be assigned to surface defects [4,13].

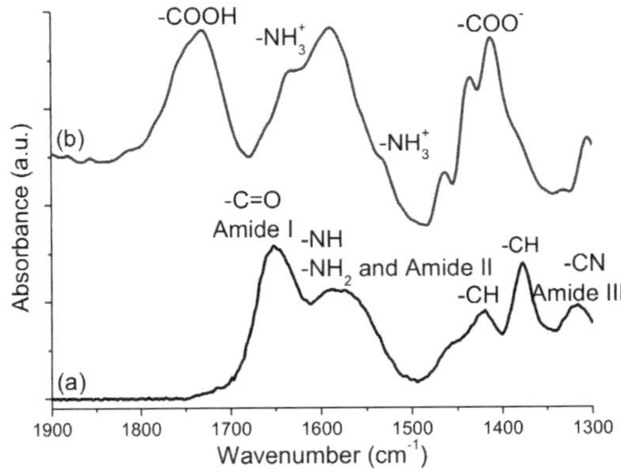

Figure 1. The FTIR spectra of chitosan (a) and CMC (b).

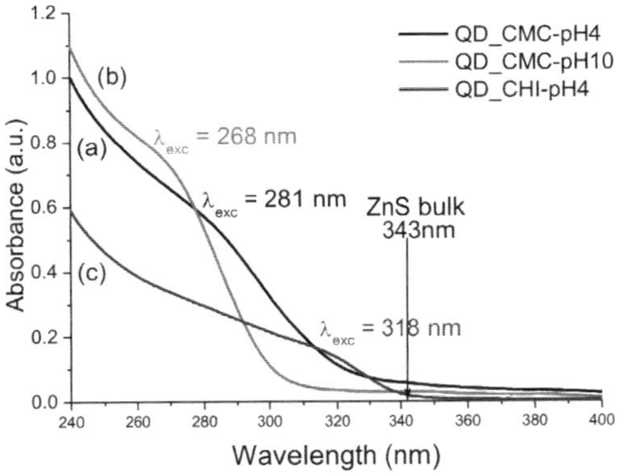

Figure 2. UV-Vis spectra for (a) QD_CMC-pH4, (b) QD_CMC-pH10, and (c) QD_CHI-pH4.

Figure 4 shows representative images of ZnS QDs produced with the CMC at pH=4.0 (A) and pH=10.0 (B) and with CHI at pH=4.0 (C) with similar morphological aspects and spherical shape. The histogram of the size distribution of QD_CMC-pH4 nanoconjugates indicates a monodisperse distribution with an average size of 4.3±0.3 nm. Correspondingly, the QD_CMC-pH10 samples exhibited reasonably monodisperse nanoparticles, with an average size cent red at 2.7±0.2 nm. The reference ZnS QDs conjugated with CHI (QD_CHI-pH4) presented an average size 5.1±0.3 nm which is relatively bigger than CMC nanoconjugates. So, the TEM results evidences that ZnS QDs were properly stabilized by CHI and CMC, in reasonable agreement with the values obtained based on the UV-vis results in the previous section.

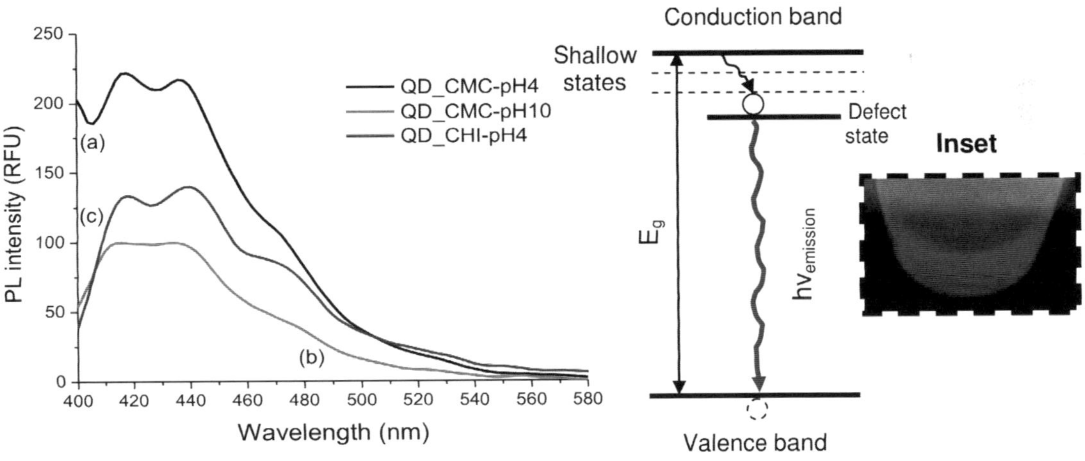

Figure 3. PL spectra of ZnS conjugates. Inset: schematic drawing of excitonic recombination pathway (not to scale) and blue luminescence image of ZnS conjugates.

4. Conclusions

In the present work, ZnS QDs directly bio-functionalized by carboxymethyl chitosan were synthesized using an eco-friendly one-step colloidal process in aqueous medium at room temperature. The results indicated that by varying the pH within a relatively broad range, from acid (pH=4.0) to alkaline

IOP Conf. Series: Materials Science and Engineering **76** (2015) 012005 doi:10.1088/1757-899X/76/1/012005

(pH=10.0) CMC solutions, the average size of ZnS QDs were significantly affected, leading to important changes in the photoluminescent properties of the ZnS nanoconjugates.

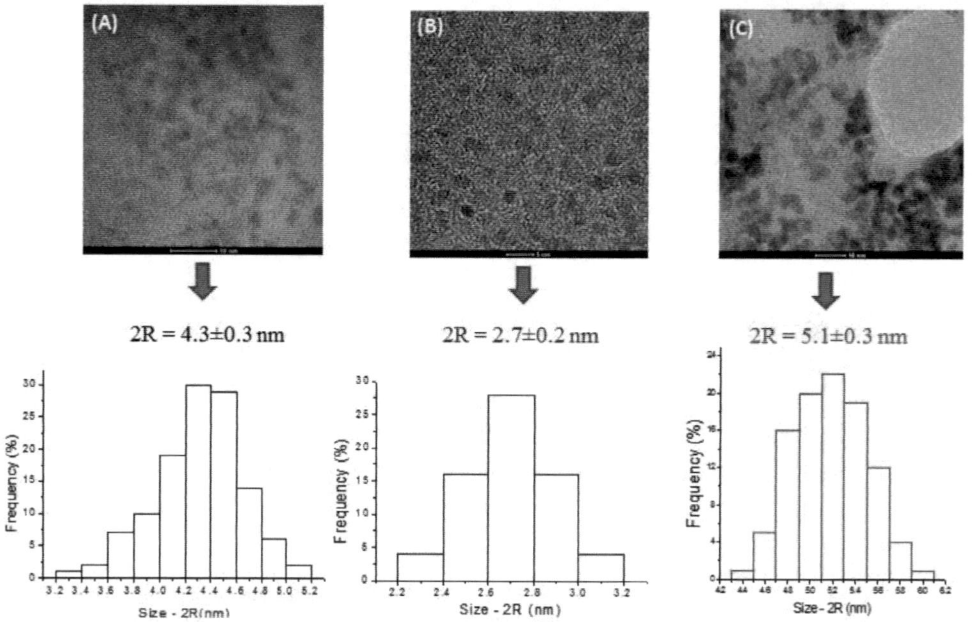

Figure 4. TEM images of QD_CMC-pH4 (A), QD_CMC-pH10 (B), and QD_CHI-pH4 (C) with their respective histograms of size distributions of ZnS nanoparticles (bottom).

Moreover, the results indicated that the stabilization mechanism of the QD_CMC nanoconjugates was determined by the interactions of the carboxylate functional groups of CMC with Zn^{2+} species at the surface of the QDs as compared to chitosan (QD_CHI) that was predominately by amine groups.

Acknowledgement
The authors acknowledge the financial support of Brazilian agencies CAPES, FAPEMIG, and CNPq. Also, they express their gratitude to the staff of the Microscopy Center/UFMG for TEM analysis.

References

[1] Etheridge M L, Campbell S A, Erdman A G, Haynes C L, Wolf S M and McCullough J 2013 *Nanomedicine* **9** 1
[2] Mansur H S, Mansur A A P, Curti E and de Almeida M V 2012 *Carbohydr. Polym.* **89** 1
[3] Mansur A A P, Mansur H S, Ramanery F P, Oliveira L C A and Souza P 2014 Appl. Catal. B: Environ. **158-159** 269
[4] Mansur H S, Mansur A A P, Ramanery F P and Borsagli F G L M 2014 J. Nanopart. Res. **16** 1
[5] Mansur H S, Mansur A A P, de Almeida M V and Curti E 2013 *J. Mater. Chem.* **1** 1696
[6] Chen X -G and Park H -J 2003 *Carbohydr. Polym.* **53** 355
[7] Jiang R, Zhu H, Yao J, Fu Y and Guan Y 2012 *Appl. Surf. Sci.* **258** 3513
[8] Abreu F R and Campana-Filho S P 2009 *Carbohydr. Polym.* **75** 214
[9] Brus L E 1984 *J. Chem. Phys.* **80** 4403
[10] Tauc J and Menth A 1972 *J. Non-Cryst. Solids* **8-10** 569
[11] Jaiswal A, Sanpui P, Chattopadhyay A and Ghosh S S 2011 *Plasmonics* **6** 125
[12] Chang S -Q, Kang B, Dai Y -D, Zhang H –X and Chen D 2011 *Nanoscale Res. Lett.* **6** 591
[13] Denzler D, Olschewski M and Sattler K 1998 *J. Appl. Phys.* **84** 2841

IOP Conf. Series: Materials Science and Engineering **76** (2015) 012006 doi:10.1088/1757-899X/76/1/012006

Photoluminescence of the Eu-doped thin film heterojunction GaAs/SnO_2 and rare-earth doping distribution

C F Bueno [1], **L V A Scalvi** [1], **M J Saeki** [2], **M S Li** [3]
[1] Physics Dept. – FC and POSMAT, UNESP São Paulo State University Bauru, SP, Brazil
[2] Chemistry and Biochemistry Dept.- IBB, UNESP, Botucatu, SP, Brazil.
[3] Institute of Physics of S. Carlos, USP University of São Paulo, São Carlos SP, Brazil

E-mail: scalvi@fc.unesp.br

Abstract. Tin dioxide (SnO_2) thin films doped with Eu^{3+}, are deposited by the sol-gel-dip-coating process on top of GaAs films, which is deposited by resistive evaporation on glass substrate. This heterojunction assembly presents luminescence from the rare-earth ion, unlike the SnO_2 deposition directly on a glass substrate, where emissions from the Eu^{3+} transitions are absent. The Eu^{3+} transitions are clearly identified and are similar to the observation on SnO_2 pressed powder (pellets), thermally treated at much higher temperatures. However, in the form of heterojunction films, the Eu emission comes along a broad band, located at higher energy compared to Eu^{3+} transitions, which is blue-shifted as the thermal annealing temperature increases. The size of nanocrystallites points toward quantum confinement or electron transfer between oxygen vacancies, originated from the disorder in the material, and trivalent rare-earth ions, which present acceptor-like character in this matrix. This electron transfer may relax for higher temperatures in the case of pellets, and the broad band is eliminated.

1.Introduction

The interest for rare-earth ions has been grown considerably in the past years, due to their optical emission lines in a large wavelength range [1], with considerable technological interest towards electroluminescent devices construction [2]. Doping semiconductors with rare-earth ions contributes to innovation in optoelectronics and, to make the emission useful for devices application, the ion must be incorporated into a matrix with adequate format, allowing the coupling to the circuit. The format of thin films is rather convenient in this direction because it is easy to link to an electronic circuit through electrical connections, such as deposited metallic contacts.

The ion Eu^{3+} present emission lines in the red region, mainly about 595nm and 615nm, which correspond to dipole magnetic transition ($^5D_0 \rightarrow {}^7F_1$) and electric dipole transition ($^5D_0 \rightarrow {}^7F_2$), respectively [3]. Tin dioxide, doped with Eu, exhibits outstanding properties of cathode luminescence, thermoluminescence and photoluminescence [4]. This semiconductor oxide has a wide bandgap (3.6-4.0eV), and has been investigated showing high applicability in many areas, such as gas sensors [5] and solar cells [6]. It presents n-type conductivity due to oxygen vacancies and tin interstitial atoms [7]. Gallium arsenide (GaAs) is an important semiconductor for the microelectronic industry, due to high mobility and direct bandgap transition [8], allowing the interaction between light and electricity, being used in heterostructures [9], light emitting diodes [10], lasers [11], among other applications.

The formation of heterojunctions, using transparent oxides deposited on top of semiconductors has

IOP Conf. Series: Materials Science and Engineering **76** (2015) 012006 doi:10.1088/1757-899X/76/1/012006

unique performances concerning the emission efficiency and optical characteristics, crucial points to the operation of functional devices [12]. A heterojunction ZnS-CdS showed properties modulated by the semiconductor ZnS [13]. Investigation on the heterojunction $SrTiO_3/Bi_2O_3$ has shown higher efficiency in the charge separation when compared to separated $SrTiO_3$ or Bi_2O_3 [14]. ZnO films were deposited on boron-doped silicon substrate (p^- e p^+), showing rectifying behavior for the heterojunction ZnO/p^--Si, whereas ZnO/p^+-Si has exhibited electroluminescence [15]. ZnO films has also been grown on GaAs substrate, forming a heterojunction with rectifying characteristics, with violet-blue and infrared electroluminescence [16], adequate for optical fiber communications.

The heterojunction SnO_2/GaAs in the form of thin films, have been successfully obtained, giving birth to smooth interface and improved electrical properties [17,18]. In this paper, luminescence from Eu-doped SnO_2 films, deposited on top of GaAs as a heterojunction, is reported. The Eu^{3+} transitions are clearly identified in the PL spectra, unlike the films of SnO_2:Eu alone, annealed at similar temperatures as the heterojunction. A broad band also shows up, which is blue shifted as the thermal annealing temperature increases, being destroyed for pressed powder, annealed to higher temperature.

2.Experimental

GaAs thin films are deposited by the resistive evaporation technique using a tungsten boat, and controlling the electrical current through the boat, under low pressure (10^{-5} mbar). After evaporation, films where submitted to annealing at 150ºC by 30 min. The deposition of Eu^{3+} - doped SnO_2 layer has been described in details elsewhere [19]. The deposition takes place in air atmosphere, and after each layer, samples are dried in air at 200ºC by 10 min. This procedure was repeated 10 times. The final annealing was at varying temperature (300 to 500°C) by 1 h. The xerogel (powder) SnO_2:2%Eu was obtained by evaporating the solvent, and treating at 1000°C by 1h. The remaining powder was pressed down to pellet format with 5000kgf/cm^2 by 3 min, and treated again at 1000°C by 20min.

X-ray diffraction data were obtained using a Rigaku diffractometer, model D/MAX 2100PC operated with a Cu K_α radiation source (1.5405 Å) and a Ni filter for reducing the undesirable K_β radiation. The data are collected with 0.02° of step and scanning rate of 1°/min in the 2theta mode for films (fixed incident angle of 1.5°) in the angular range of $20^o \leq 2\theta \leq 80^o$.

For the photoluminescence (PL) measurements, samples were excited with a modulated 350 nm line of a Kr^+ laser and the signal was detected by a R955 PMT from Hamamatsu and a SR530 Lock-in Amplifier from Stanford Research System. A single configuration monochromator was used for scanning the PL spectra. Morphological characterization and compositional analysis were carried out with a scanning electron microscopy (SEM) system (FEI model Quanta 200), equipped with an energy dispersive X-ray (EDX) microanalysis system (Oxford model Inca 250P20).

3. Results and Discussion

Figure 1 shows the PL of the heterojunction with distinct thermal annealing. The inset of figure 1 is the same measurement carried out for a sample of SnO_2:2at%Eu, in the format of pellet (pressed powder). Three well defined peaks are identified for the heterojunction samples and are coincident with Eu^{3+} transitions: $^5D_0\rightarrow{}^7F_1$ (about 596nm), $^5D_0\rightarrow{}^7F_2$ (about 618nm) and $^5D_0\rightarrow{}^7F_4$ (about 704nm). Besides, a broad band is also identified, with peak about 550 nm for the sample treated at 300ºC, and blue-shifted for the samples treated at higher temperatures. The peak is about 470nm for the sample treated at 500°C. The emission spectra of SnO_2:2at%Eu film, not coupled to a GaAs layer on a heterojunction, presents only the broad band, not allowing the identification of the Eu^{3+} transitions. Concerning the transition identified for the powder SnO_2 samples (inset of figure 1), they are easily identified and take place on a wider range. No broad band is obtained for the powder. Eu^{3+} peaks are seen at 557, 565 and 572 nm corresponding to the transition $^5D_1\rightarrow{}^7F_2$, 583 nm corresponding to $^5D_1\rightarrow{}^7F_3$, 593, 598 and 604 nm ($^5D_0\rightarrow{}^7F_1$), 614, 618 and 625 nm ($^5D_0\rightarrow{}^7F_2$), 633, 638 e 657 nm ($^5D_0\rightarrow{}^7F_3$), and 716, 721 and 727nm ($^5D_0\rightarrow{}^7F_6$). Peaks related to $^5D_0\rightarrow{}^7F_1$ transition are the most intense, unlike the PL for the heterojunction films shown in fig. 1, where the transition $^5D_o\rightarrow{}^7F_2$ turned out as the most intense. It is also different from excitation with the fourth harmonic of a Nd:YAG laser (266 nm) for the same level of doping concentration, where the transition $^5D_0 \rightarrow ^7F_2$ is the most intense [19]. The dipole magnetic transition $^5D_0 \rightarrow ^7F_1$ is related to Eu^{3+} ion substitutional to Sn^{4+} in the SnO_2 matrix sites whereas radiation from the electric dipole transition $^5D_0 \rightarrow ^7F_2$ is mainly due to ions located at particles surface. Then, it is expected that the former transition be more evident for the films where the crystallites are smaller compared to the pellets and then, the depletion layer regions are wider.

For the heterojunction films, the Eu emission comes along a broad band which is blue- shifted as the thermal annealing temperature increases. It is well-known that increasing the annealing temperature, the crystallite size must increase. Although quantum confinement seems to be an explanation, this evidence is in opposition to what would be expected, since the dependence on the crystallite size is similar to quantum dots. Then, for a particle localized in a nanocrystallite, as the size decreases, the emission spectra should shift to high energies [20,21], in opposition to what was observed here. Another possible contribution for the detected broad band is related to electron transfer between oxygen vacancies and acceptor Eu^{3+} ions. These phenomena relax for annealing at higher temperatures and longer times, in the case of pellets, and eliminate the broad band from the PL spectra. Reported results for Sb-doped SnO_2 samples ascribe the main origin of the UV-violet luminescence band to the electron transition between the donor level and the acceptor level formed by Sb ions [22]. Considering that it is much more common to a trivalent rare-earth ion, such as Eu^{3+}, to form an acceptor level in SnO_2, than a Sb ion, whose predominant state is pentavalent [23], it is expected that the broad band seen in Fig. 1 has contribution related to the electron transition between the donor level, formed by oxygen vacancies, and the acceptor level formed by Eu^{3+} ions. It also leads to high charge compensation as determined by electrical characterization measurements [24].

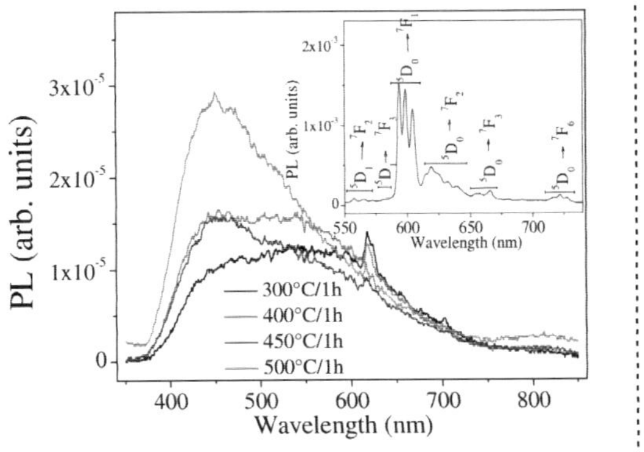

Figure 1. PL of the heterojunction with distinct thermal annealing. Inset: PL of sample SnO_2:2at%Eu, in the format of pressed powder (pellet)

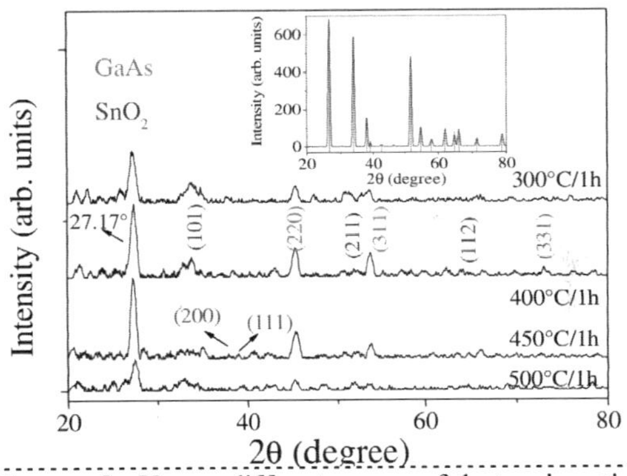

Figure 2. X-ray diffractogram of heterojunctions. Annealing temperatures are indicated and main directions are labeled. Inset: diffractogram of theSnO_2:2%Eu powder. Red vertical lines are from the reference JCPDS- 41-1445 for tin dioxide..

Figure 2 shows X-ray difratograms for these samples, which exhibit the noisy and diffuse profile typical of materials composed of nanocrystalline building blocks. The labeled peaks correspond to the planes of rutile structure of tin dioxide (JCPDS- 41-1445) and planes of GaAs (JCPDS-80-0016). Characteristic peaks of SnO_2 are seen at 33.7° and 51.5° corresponding to (101) and (211) crystallographic planes, and characteristic peaks of GaAs at 45.4° and 53.7°, related to (220) and (311) planes. At 27.17° there is a probable overlapping of two peaks: one related to plane (100) of SnO_2, and another peak corresponding to plane (111) of GaAs. Concerning the diffractogram of the powder sample, it exhibits many peaks related to the rutile SnO_2, as can be seen in the inset of fig. 2, by the very good match with the red lines of the JCPDS- 41-1445 reference. Crystallite size evaluated according to the Scherer equation in the (101) SnO_2 direction grows from 11.3nm for heterojunction sample annealed at 300°C to 15.6nm for annealing at 500°C, and 22.1 for the powder. It confirms the expectation of crystallize growing with increasing annealing temperature.

Figure 3 (a) is a SEM image of a heterojunction GaAs/SnO_2:2at%Eu, annealed at 400°C by 20min. Figure 3(b) is the EDX spectra for the same sample, in the region indicated by the yellow arrow in the figure 3(a). The SEM image shows some prominences, which means most external particles. The EDX analysis carried out the sample surface identified variation of up to 700% in the relative Eu/Sn composition, for these prominent rounded regions. Although it is not possible to ensure that electrons interacted only with the elements within the demarked area under this magnification and so high voltage, these regions seem to be very important in this variation. The EDX spectra shown in figure 3(b) clearly confirm the Eu peaks. The inset of figure 3(b) shows an EDX general field of the marked region (squared) of the figure 3(a), showing an illustration of the Eu distribution on the sample surface. Although the Eu ions are spread over the whole surface, there are clearly regions where they are more concentrated. These prominent regions were not identified for samples where the SnO_2 layer is deposited directly on the substrate, nor on heterostructures surface where the top layer is GaAs [17]. Then, we believe that these regions play a fundamental rule in order to concentrate the Eu^{3+} ions and provide luminescence from them.

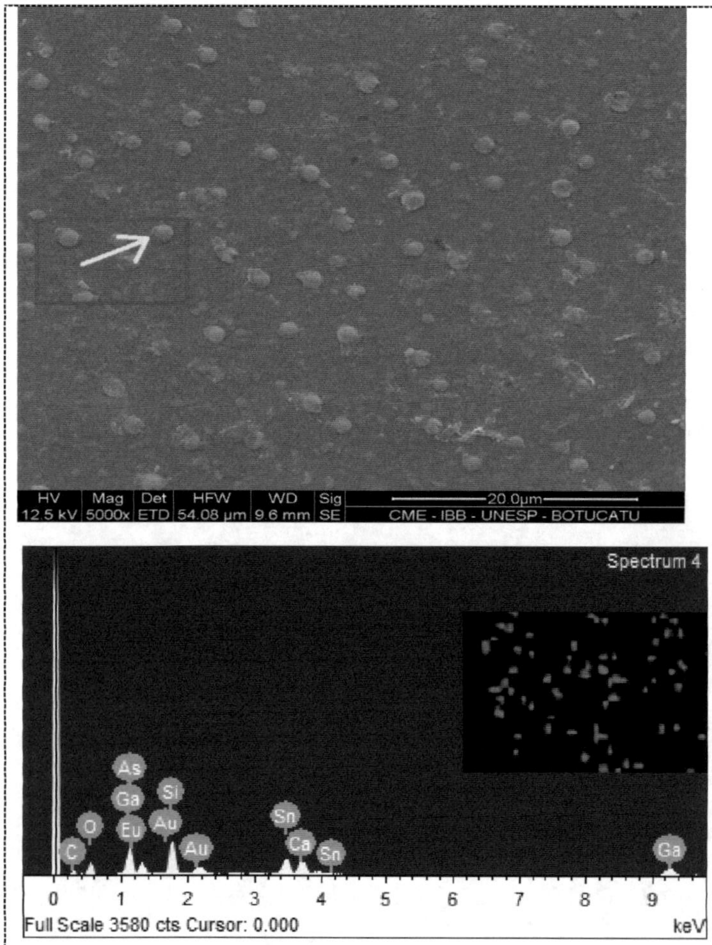

Figure 3. (top) MEV of the heterojunction GaAs/SnO_2 where the top layer is annealed at 400°C by 20min, magnitude 5000x: **(bottom)** EDX spectra for the same material in the region indicated by the red arrow in the top figure. **Inset** :EDX general field of the heterojunction GaAs/SnO_2:2%Eu treated at 400°C/20min, showing the Eu distribution.

4. Conclusion

Main transitions from Eu^{3+} where identified in the matrix SnO_2 in a heterojunction with GaAs, unlike films of sole SnO_2 with the same doping. It comes along a broad band, blue shifted for larger crystallites

The luminescence from Eu in thin films is a rather convenient way, providing accessibility to construction and operation of luminescent and electroluminescent devices. The combination with GaAs in the form of heterojunction films provides the red emission, not seen from Eu-doped films deposited in an isolated way, when thermally annealed at same temperature range.

Acknowledgements

We thank the financial support of the Brazilian agencies: CNPq and FAPESP.

References

[1] Eliseeva S V and Büznli J-C G 2010 *Chem. Soc. Rev.* **39** 189
[2] Perea-Lopez N, Gonzalez-Ortega J A and Hirata G A 2006 *Opt. Mat.* **29** 43
[3] Lahiri D, Ningthoujam R S, Bhattacharyya D and Sharma S M 2010 *J. Appl. Phys.* **107** 054316
[4] Daboussi S, Elhouichet H, Ajlani H, Moadhen A, Oueslati M and Roger J 2006 *J. Lumin.* **121** 507
[5] Pourfayaz F, Mortazavi Y, Khodadadi A and Ajami S 2008 *Sensors Actuators B: Chem.* **130** 625
[6] Chou S-M, Hon M-H, Leu I-C and Lee Y-H 2008 *J. Elect. Soc.* **155** H923
[7] Watanabe K, Hashiguchi M, Sakaguchi I, Bryant A, Adachi Y, Zhen Y, Ohgaki T, Ohsawa T, Haneda H and Ohashi N 2014 *Appl. Phys. Lett.* **104** 042110
[8] Liu N and Kuech T F 2005 *J. Elect. Mat.* **34** 1010
[9] Lodha S and Janes D B 2006 *J. Appl. Phys.* **100** 1010
[10] Zhao Z-Y, Song Z-Q, Shi W-Z and Zhao Q-Z 2014 *Opt. Exp.* **22** 11654
[11] Yoo B-S, Chu H Y, Park H-H, Lee H G and Lee J 1997 *IEEE J. Quant. Elect.* **33** 1794
[12] Tian W, Zhang C, Zhai T, Li S-L, Wang X, Liu J, Jie X, Liu D, Liao M, Koide Y, Golberg D and Bando Y 2014 *Adv. Mat.* **26** 3088
[13] Hernández-Gordillo A, Tzompantzi F, Gómez R and Cálderon-Benavides H 2014 *Mat. Lett.* **115** 147
[14] Zhang Z, Zhu J, Li S and Mao Y 2014 *J. Sol. Sta. Chem.* **211** 120
[15] Chen P, Ma X and Yang D 2007 *J. Appl. Phys.* **101** 053103
[16] Du G, Cui Y, Xiaochuan X, Li X, Zhu H, Zhang B, Zhang Y and Ma Y 2007 *Appl. Phys. Lett.* **90** 243504
[17] Pineiz T F, Scalvi L V A, Saeki M J and Morais E A 2010 *J. Electron. Mat.* **39** 1170
[18] Pineiz T F, Morais E A, Scalvi L V A and Bueno C F 2013 *Appl. Surf. Sci.* **267** 200
[19] Morais E A, Scalvi L V A, Tabata A, Oliveira J B B and Ribeiro S J L 2008 *J. Mat. Sci.* **43** 345349
[20] Alivisatos A P 1996 *Sci.* **271** 933
[21] Ledoux G, Gong J, Huisken F, Guilois O and Reynaud C 2002 *Appl. Phys. Lett.* **80** 4834
[22] Wang Y, Ma J, Si F, Yu X and Ma H 2005 *J. Lumin.* **114** 71
[23] Geraldo V, Briois V, Scalvi L V A and Santilli C V 2010 *J. Phys. Chem.* C **114** 19206
[24] Morais E A, Scalvi L V A, Cavalheiro A A, Tabata A and Oliveira J B B 2008 *J. Non-Cryst. Sol.* **354** 4840

IOP Conf. Series: Materials Science and Engineering **76** (2015) 012007 doi:10.1088/1757-899X/76/1/012007

Characterization of graphene oxide nanofilms obtained by the SAW atomization

O V Balachova[1,4], S M Balashov[1], L T Kubota[2], R A Timm[2], P H Nascimento[1], A Pavani Filho[1] and S Moshkalev[3]

[1] Department of Microsystems, Center for Information Technology Renato Archer, Rod. D. Pedro I (SP-65), km 143.6, Amarais, Campinas-SP, 13069-901, Brazil
[2] Department of Analytical Chemistry, State University of Campinas, Campinas-SP, 13083-970, PO box 6154, Brazil
[3] Center for Semiconductor Components, State University of Campinas, r. João Pandia Calógeras, 90, Campinas-SP, 13083-870, PO box 6101, Brazil

E-mail: olga.balachova@cti.gov.br

Abstract. Due to its ability to absorb water molecules, graphene oxide (GO) is considered a promising material for sensitive coatings in fast surface acoustic wave (SAW) humidity sensors. In this work, we characterize GO films obtained by the SAW atomization technique. It is shown that the atomized submicroliter droplets of aqueous suspension of GO can be deposited onto the surface of Si, $LiNbO_3$ or quartz substrates forming discrete or continuous films of nanometer thickness. The deposited films were examined using AFM and electron microscopy. We discuss the dependence of thickness and structure of the obtained GO films on the parameters of deposition: the number of atomized droplets, a volume of the initial droplet, a distance between the atomizer and the sample, etc. To evaluate the adsorption characteristics of the obtained GO films, we used them as sensitive coatings of the SAW humidity sensors. We found that the adsorption characteristics of the GO films are determined by fast adsorption on the surface of GO sheets and slow adsorption, attributed to limited penetration of water molecules between the sheets, and depend on the number of deposited layers.

1. Introduction

Surface acoustic wave (SAW) sensors can be used for monitoring various chemical and physical parameters: strain, atmospheric pressure, humidity, biological matter, etc. [1-5]. The operating principle of these devices is based on a high sensitivity of the SAWs to the condition of the surface on which they propagate. For that reason some types of the SAW sensors–gas, biological, or humidity sensors–must have special coatings deposited onto the SAW propagation path between the interdigital transducers (IDTs) of the device. For example, thin films of polyvinyl alcohol (PVA) [2,3] and graphene oxide (GO) [4] find use as sensitive coatings for humidity sensors.

Thin GO films can be obtained from aqueous dispersions of GO with subsequent water evaporation by spin coating, spray coating or dip coating [6,7]. All these methods use large volumes of suspensions and have their limitations: spin coating, for example, cannot be applied for rectangular substrates, or very small substrates, or for the patterned film deposition. The drawback of dip coating

[4] To whom any correspondence should be addressed

and spray coating is that the obtained films are relatively thick. Also, evaporation of thick water layers can cause a significant non-uniformity of the final thickness of a film ("coffee-ring" effect).

To obtain thin GO films, we used the SAW atomization technique [4,8]. The idea of this method is to produce a very fine spray/aerosol from a single micro or submicroliter droplet of GO aqueous suspension and deposit it onto the surface of a substrate, where it will form a very thin wet layer. After evaporation of water, GO sheets will remain at the surface forming a very thin coating. The advantage of this method is that it allows controlling a volume of the atomized suspension and, therefore, a thickness of the deposited film. Another advantage is that it allows working with microvolumes of suspensions. Also, the SAW atomization can be applied for a patterned deposition of different films onto the surfaces of various shapes and dimensions.

In this work we study the dependence of thickness and structure of thin GO films on the parameters of the SAW atomization: the number of atomized droplets, a volume of the initial droplet, a distance between the atomizer and the sample, the applied RF power and sample surface preparation. Also, we evaluated the adsorption characteristics of the obtained GO films using them as humidity-sensitive coatings in the SAW sensors.

2. Experimental procedure

The graphene oxide films were deposited onto the surface of quartz, $LiNbO_3$ and Si substrates by atomization of 0.2 and 0.35-μl droplets of GO aqueous dispersions by the SAW atomization technique.

2.1. GO aqueous suspensions

First, graphite oxide was obtained from natural graphite by the modified Hummers' method. In our case, graphite (3.0 g) and $NaNO_3$ (1.5 g) were mixed with concentrated H_2SO_4 (69 ml) and cooled in an ice bath. A small amount of $KMnO_4$ was added to the obtained suspension. The suspension was heated for 30 min, cooled in an ice bath and diluted with DI water (138 ml, 18 MΩ·cm). Then, the suspension was left to react for 15 min at 98° C, cooled again and mixed with DI water (420 ml) and 30% H_2O_2 (3 ml). The final product was washed with H_2O, HCl and C_2H_5OH, then filtered and dried. The obtained material was dispersed in DI water (1 mg/ml) and exfoliated to GO in ultrasonic bath for 1 h. The obtained dispersion was diluted with DI water (1:1) and centrifuged at 1700 x g for 10 min. Immediately after centrifugation the supernatant was carefully removed with a pipette and transferred to a clean glass container. This procedure allows obtaining stable homogeneous GO dispersions that can be stored for few weeks without changing their properties.

2.2. SAW atomization

The SAW atomizing module is shown in the figure 1.

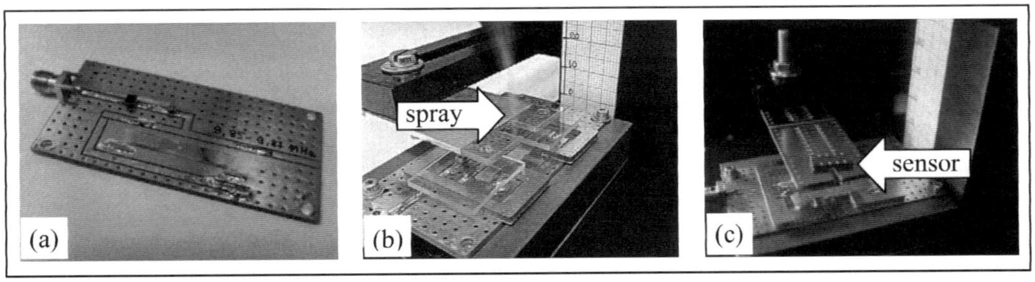

Figure 1. The atomizing module: SAW atomizer (a) in operation (b); the SAW humidity sensor (open active surface down) above the atomizer during atomization.

In this work we used the atomizers with symmetric IDTs fabricated on 128°-rotated Y-cut of $LiNbO_3$ by a conventional photolithography. Central frequency of the devices was 9.85 MHz. To atomize one droplet of GO suspension from the center of the atomizer, the RF signal was applied to the IDTs. The obtained aerosol was deposited onto the sample above the atomizer forming a thin coating. The atomization process took 1-2 s and water evaporation took about a minute. More detailed description of the SAW atomization technique can be found elsewhere [4,7].

3. Results and discussion

Although the GO aerosol obtained by the SAW atomization of submicroliter droplet can spread on large areas (up to 15 cm^2), the resulting film is not uniform (figure 2a): its thickness decreases noticeably from the center to borders (figures 2b and 2c). Only a central part of the film (in our case, it is a circle of 2.5 mm in diameter, approximately) can be considered uniform. This region is situated directly above the center of the atomizer during deposition and obtains a maximum amount of graphene oxide with aerosol.

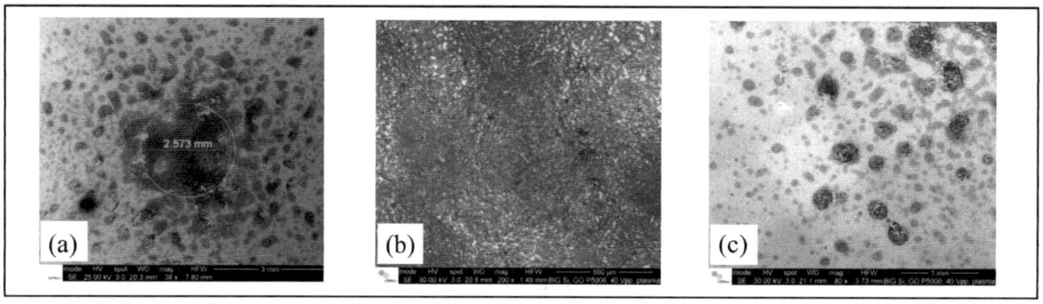

Figure 2. GO film on the Si surface: general view (a), the central spot (b) and the border (c).

The dependence of the thickness of GO films on the number and volume of atomized droplets is shown in the figure 3 and figure 4. It can be seen that thicker films are obtained by increasing the number of atomizations and/or by atomizing larger droplets, keeping constant the other parameters of the process. Adjusting the parameters of the process, it is possible to obtain continuous uniform coatings.

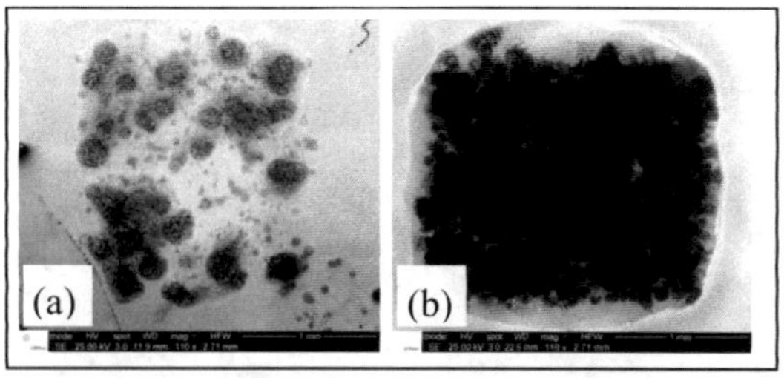

Figure 3. GO films at the silicon: one (a) and ten (b) depositions of 0.2-μl droplets.

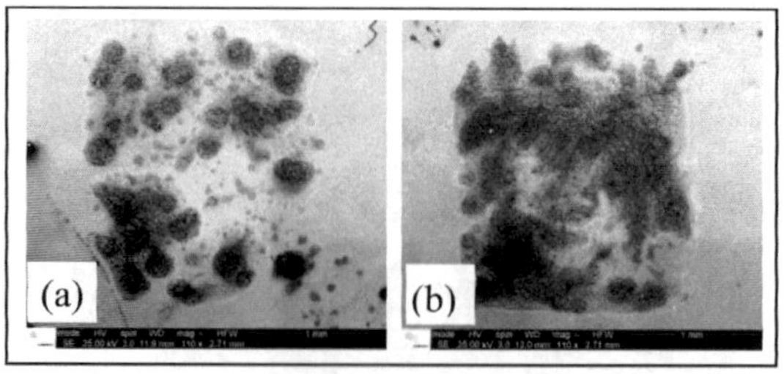

Figure 4. GO films at the Si: one deposition of 0.2-μl (a) and 0.35-μl (b) droplets.

Another parameter that influences the quality of the film is the distance between the atomizer and a sample: more uniform films were obtained at 6-9 mm from the atomizer surface. Also, thicker GO films with higher degree of surface coverage were obtained at higher RF power for the same parameters of the process (figure 5).

We observed more uniform coatings and higher degree of coverage at the surfaces activated in oxygen plasma before the film deposition (15 min at 100 W and 200 mTorr) for all substrates (figure 6). SEM images demonstrated that after one or two depositions a substantial area of the coatings is formed by single or double GO sheets (figure 7). AFM measurements showed that the thickness of the films in this case is about 2.5-3 nm. The films deposited onto the non-activated by plasma samples (figure 6a) are much thicker: 40-100 nm after one deposition. It results from the fact that activated surfaces ensure better spreading of the deposited spray droplets before drying.

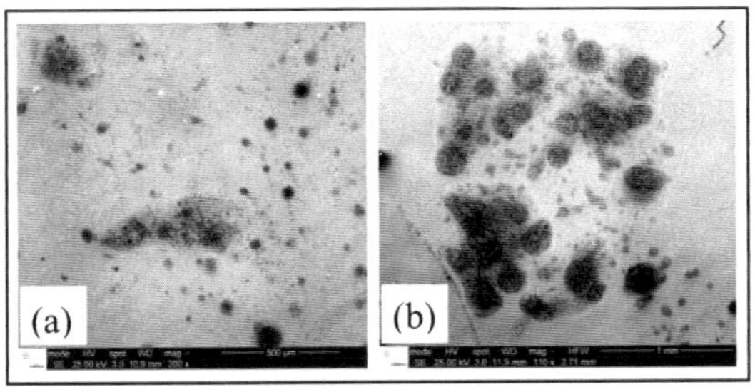

Figure 5. GO films on the Si surface: 1 W (a) and 4 W (b) of the applied power.

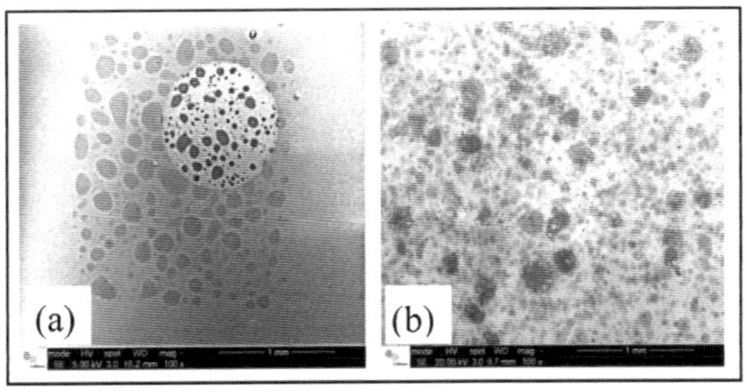

Figure 6. GO films on $LiNbO_3$: no surface activation (a) and after surface activation (b).

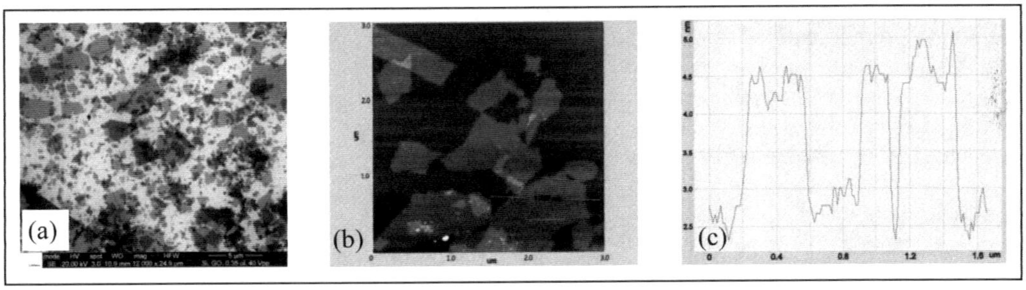

Figure 7. GO film: SEM image (a), AFM image (b) and thickness measurements (c).

4. Adsorption characteristics of GO films

To evaluate the adsorption characteristics of the obtained GO films, we used them as sensitive coatings of the SAW humidity sensors. We monitored the kinetics of evaporation of water molecules from the surface of the liquid sample of saturated salt into the enclosed chamber with initially zero relative humidity. The details of these experiments can be found elsewhere [4,9]. Typical experimental differential frequency response of the sensor $f(t)$, shown in the figure 8, was approximated by the multi-exponent equation, which corresponds to N adsorption processes

$$f(t) = F_0 + \sum_{i=1}^{N} A_i(1 - \exp(-\alpha_i t)) \quad (1)$$

where F_0 is a baseline differential frequency, which corresponds to the response of the sensor at the initial moment, when relative humidity is equal to zero; A_i and α_i are the correspondent saturation amplitude and rate coefficient [10,11]. The results of LSQ-fitting of typical experimental data with (1) for N=1 and N=2 (figure 8) demonstrate that the approximation with N=2 is more appropriate. The values of A_i and α_i (i=1,2) for a wide range of humidity are presented elsewhere [4]. The analysis of these data shows that the saturation time (rate coefficient) of the first process is approximately ten times smaller (bigger) then of the second one. The first process (i=1) can be attributed to the process of evaporation of water molecules from the surface of the liquid sample and their adsorption by the sites on the surface of the GO sheets. These sites have the lowest adsorption energy and, hence, the smallest adsorption time. The second process (i=2) can be attributed to the adsorption on the border of GO sheets or between them. The response time of the sensor grows with the number of deposited films due to a higher degree of surface coverage by GO and a larger number of adsorption sites.

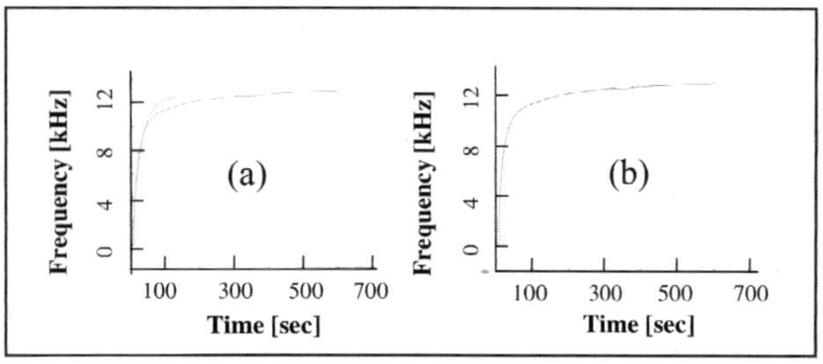

Figure 8. Typical experimental kinetic adsorption curve (blue) of the GO SAW humidity sensor. Red curves represent its LSQ-fitting results with (1) for N=1 (figure 8a) and N=2 (figure 8b).

5. Conclusions

The SAW atomization technique can be used for patterned deposition of GO films from aqueous GO suspensions onto the surface of different substrates including quartz used in the SAW humidity sensors. The films form discrete or continuous coatings at the surface of the substrates/sensors with the minimum thickness of 2.5-3 nm which corresponds to a GO monolayer. The thickness of the deposited films strongly depends on the number of atomized droplets, a volume of the initial droplet, a distance between the atomizer and the sample, the applied RF power and sample surface preparation and can be controlled. Discrete coatings can be obtained by a small number of the atomized droplets, but to obtain the continuous coatings, it is necessary to make at least 10 depositions. The thickness of the films remains constant in its central area of 2.5 mm in diameter approximately. Water adsorption is determined by the fast process on the surface of GO sheets and the slow process attributed to the limited adsorption between GO flakes and grows with the number of deposited layers.

Acknowledgements

We thank the INCT NAMITEC (FAPESP pr. 08/57862-6), CNPq (pr. 590032/2011-9 and pr. 302229/2013-2) for financial support and C2NANO–Center for Nanoscience and Nanotechnology/ MCT for the privilege of using a scanning electron microscope SEM-Inspect at the Electron Microscope Laboratory–LME. We are grateful to M. A. Canesqui, A. V. U. Braga and E. F. Z. A. von Zuben for their collaboration.

References

[1] Konno R, Mitsui M, Kuwano H, Nagasawa S, Sano K and Hagasaka J **2007** A highly sensitive strain sensor using surface acoustic wave and its evaluation for wireless battery-less sensor network *2007 IEEE Sensors Conf.: Proc. (Atlanta, GA, USA, 28-31 October 2007)* pp 796-799

[2] Jiang Q, Yang X M, Zhou H G and Yang J S 2005 *Sensors and Actuators* A **118** 1-5

[3] Buvailo A, Xing Y, Hines J and Borguet E 2011 *Sensors and Actuators* B **156** 444-449

[4] Balashov S M, Balachova O V, Braga A V U, Bazetto M C Q and Pavani Filho A **2013** The optimized SAW humidity sensor with nanofilms of graphene oxide *2013 IEEE Sensors Conf.: Proc. (Baltimore, MD, USA, 3-6 November 2013)*

[5] Länge K, Rapp B E and Rapp M 2008 *Anal. Bioanal. Chem.* **391** 1509-19

[6] Yao Y, Chen X, Guo H and Wu Z 2011 *Appl. Surf. Sci.* **257** 7778-82

[7] Eda G and Chhowalla M 2010 *Adv. Mater.* **22** 2392-2415

[8] Balashov S M, Pavani Filho A, Balachova O V and Almeida M G 2009 *ECS Transactions* **23** 287-293

[9] Balashov S M, Balachova O V, Braga A V U, Bazetto M C Q, Pavani Filho A and Moshkalev S **2013** Kinetic characteristics of the SAW humidity sensor partially coated with graphene oxide thin film *2013 Symp. on Microelectronics Technology and Devices (SBMicro): Proc. (Curitiba-PR, Brazil, 2-6 September 2013)*

[10] Marczewski A W 2007 *Appl. Surf. Sci.* **253** 5818-26

[11] Marczewski A W 2007 *Appl. Surf. Sci.* **256** 5145-52

Characterization of the semi-insulating properties of $AlHfO_{3.5}$ for power devices

B S Alandia[1], D R Huanca[1], V Christiano[1], S G dos Santos Filho[1]
[1]University of São Paulo: LSI/PSI/EPUSP – São Paulo, Brazil

E-mail: balandia@lsi.usp.br

Abstract. Physical and electrical characterization of hafnium aluminates gate dielectrics was carried out to investigate their semi-insulating characteristics as passivating layer for power devices. The deposited films were annealed in pure nitrogen to simulate the thermal budget during a conventional CMOS processing. C-V measurements were performed having as a result a high-frequency behavior of the flat band voltage (V_{FB}) and values greater than or equal to zero. On the other hand, the leakage phenomenon was modeled with a simplified electrical model using a leakage admittance Y_C whose influence was predominant at the accumulation region. Using X-ray reflectometry (XRR), the average thickness obtained was 15.5nm and a leakage process was inferred to occur for $AlHfO_{3.5}$ due to the observed phase separation and crystallization that occurs after annealing in pure N_2.

1. Introduction

Semi-insulating materials have been considered to replace the silicon dioxide passivation layers in power devices such as thyristors and power transistors [1-3]. An example is the semi-insulating polycristaline silicon (SIPOS) that can be used to increase the breakdown field without increasing the on-state series resistance [1-3]. In this case, the potential drop across the layer is controlled by the high volume concentration of the charged states. Two important electrical properties of the semi-insulating films are the shift of the flatband voltage due to the charging effect and the leakage process, which means a shunt parasitic resistances coexisting with a series resistance associated to the substrate [3-6]. Aluminates are possible candidates to be used as semi-insulating materials since they have bulk states at high concentrations to be charged in order to promote the increase of the breakdown field [3,7,8].

Literature points out the increase of the leakage current density for hafnium aluminates as the thermal budget of annealing is increased. This effect occurs due to the phase decomposition and crystallization of the hafnium aluminates. On the other hand, thermodynamic stability from interfacial layer leads to well-behaved electrical characteristics and low interface state density (Dit) [9]. For Capacitance-Voltage (C-V) characteristics, a decrease of the accumulation capacitance is reported due to the presence of the leakage current [9-11].

In this work, hafnium aluminates were electrically characterized according to the flatband voltage and the leakage current through the MOS structure.

2. Experimental

At first, the samples underwent an IMEC clean. $AlHfO_{3.5}$ was deposited by Atomic Layer Deposition on Si(100) substrate with equal molar concentrations of aluminum and hafnium. Layers were deposited in an ASM Pulsar 3000 connected to a Polygon 8300 Platform on Si substrate, on which a

1 nm thick chemical oxide has been grown [14]. Following, the sample was also annealed using an ASM Levitor 4300 at 1000°C during 60s in N_2 and, it was subsequently immersed into a boiling isopropyl alcohol for 1h. Finally, aluminum was deposited (200 nm thick) through a mechanical mask to define MOS capacitors with area of $2x10^{-2}cm^2$ and they were sintered at 420°C in ultrapure N_2+10%H_2 during 30min.

MOS capacitors were electrically characterized with the aid of HP 4280 C at 1MHz or Agilent E4980A Precision LCR Meter at 1MHz in order to obtain parameters from C-V characteristics, including series resistance (R_S) and effective charges in the diectrics (Q_{SS}/q). Grazing Incidence X-Ray Diffraction (GIXRD) was employed to the films using an X-ray beam with λ= 0.1542 nm at an incidence angle of 0.7°. X-ray reflectometry (XRR) was employed to obtain the physical thickness of the films. The XRR curves have a typical wavy appearance known as Kiessig fringes [13-16]. These fringes appear due to the interference effect of the beams reflected at the air/dielectric and dielectric/silicon interfaces [15,16]. The periodicity of these ondulations is related to the thickness, while the fall rate is influenced by the roughness of both interfaces [13,14,17]. Based on the periodicity of the peaks in the XRR curves, an estimation of the thickness, valid for homogeneous films, was made using the relation $T_{XRR}=2\pi/\Delta q_z$ [13,14,17]. In addition, the XRR curves were fitted using the PARRAT32 code. This computational program uses Parrat formalism [18] along with the Névot- Croce factor that takes into count the surface roughness effect upon the reflectivity [19] and the variation of the electronic density in depth. Mathematically they are expressed as:

$$R_j = \frac{r_{j,j+1} + R_{j+1}\exp(it_{j+1}k_{z,j+1})}{1 + r_{j,j+1}R_{j+1}\exp(it_{j+1}k_{z,j+1})} \quad (1)$$

$$r_{j,j+1} = \frac{k_{z,j} - k_{z,j+1}}{k_{z,j} + k_{z,j+1}}\exp(-2\sigma_{j+1}^2 k_{z,j}k_{z,j+1}) \quad (2)$$

$$k_{z,j} = \sqrt{k_{z,0}^2 - 4\pi\rho_j} \quad (3)$$

The reflectivity of the system is given by $R=|R_0|^2$, $R_{N+1}=0;\ R_N=r_{N,N+1}$. In equation (3), t_j and k_j are the thickness and wave number of the j-th layer, while ρ_j is the complex value of the electron density and σ_j is the roughness of the j-th layer.

3. Electrical Modeling

C-V curves were modeled by the electrical circuits shown in figure 1, where *Co* is the film capacitance, *Yc* is an admittance that represents a leakage process and R_S [1, 20] the series resistance associated to the silicon substrate [8].

Based on equal impedances of the electrical circuits shown in Fig. 1a and 1b, an expression for the corrected capacitance in the accumulation (C_o) and the admittance Y_c across the leakage current path are given by [20].

$$C_o = \frac{C_{ma}\left((\omega C_{ma})^2 + G_{ma}^2\right)}{(\omega C_{ma})^2 + \left(G_{ma} - R_s\left((\omega C_{ma})^2 + G_{ma}^2\right)\right)^2} \quad (4)$$

$$Y_c = \sqrt{\omega C_o\left(-\omega C_o + \omega C_{ma}\left(1 + \frac{G_{ma}^2}{(\omega C_{ma})^2}\right)\right)} \quad (5)$$

Where C_{ma} is the measured capacitance, G_{ma} is the measured conductance and $\omega = 2\pi f$. In addition, the effective charge in the dielectrics (Q_{SS}/q) is given by:

$$\frac{Q_{SS}}{q} = \frac{\left(-0.6 - \frac{kT}{q} ln\left(\frac{N_A}{n_i}\right) - V_{FB}\right)\frac{C_o}{A_G}}{q} \qquad (6)$$

Where N_A is the substrate doping, n_i is the intrinsic concentration of carriers, V_{FB} is the flatband voltage, A_g is the area of the capacitor and $kT/q = 0.0259V$ at room temperature.

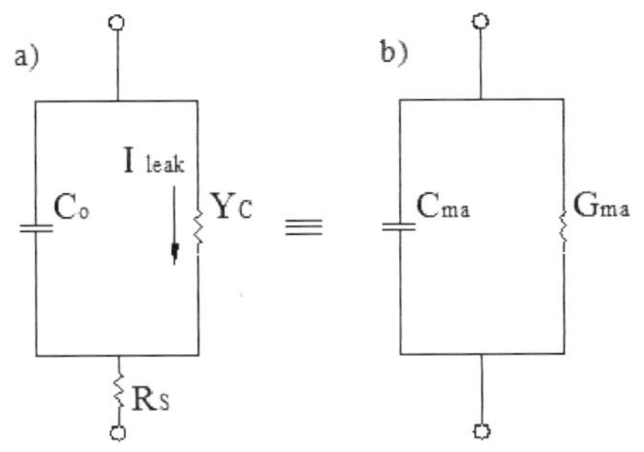

Figure 1. (a) MOS electrical model and (b) C_{ma} and G_{ma} measured by the *LCR Meter* in the accumulation.

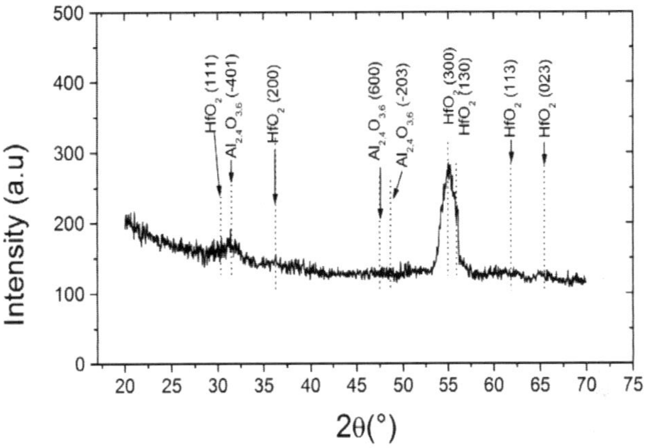

Figure 2. Typical GIXRD for the $AlHfO_{3.5}$ film.

4. Results and Discussion

Figure 2 shows the GIXRD for a typical $AlHfO_{3.5}$.film. It is noteworthy that, after annealing in N_2 by RTP, the matrix film passes to contain both crystalline $Al_{2.4}O_{3.6}$ and HfO_2 phases, indicating phase separation and crystallization.

X-ray reflectometry (XRR) was employed to obtain the physical thickness of the films. Although the films were supposed to be single layer, they could not be fitted using the single layer model, being necessary a four-layer one. The XRR curve shown in figure 3 was fitted with the aid of the PARRAT32 code using the procedure mentioned in the experimental. The $AlHfO_{3.5}$ film was 15.5nm in physical thickness and it was slightly different of that computed using the simplified equation $T_{XRR}=2\pi/\Delta q_z$. The surface roughness of these films was lower than 0.6nm.

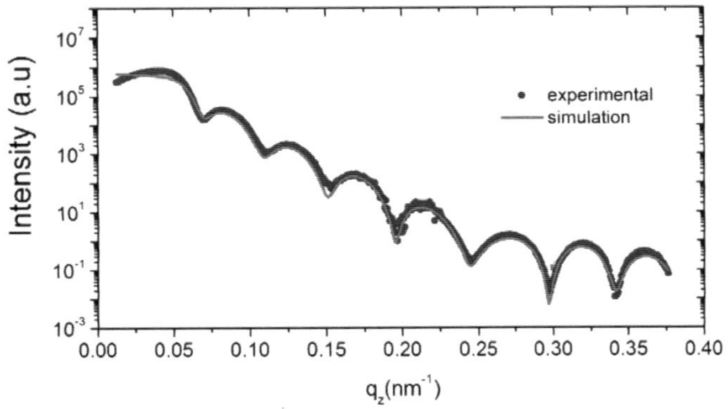

Figure 3. Typical XRR for the $AlHfO_{3.5}$ film and the fitting.

As already mentioned, for the $AlHfO_{3.5}$, phase separation and cristallization takes place, promoting the formation of HfO_2 and $Al_{2.4}\,O_{3.6}$ crystalline phases. Also, non-homogeneous electronic density distribution in the dielectrics gave rise to the formation of several regions which can be viewed as fitting sub-layers, each of them with arbitrary electron density to obtain the real thickness using the PARRAT32 code.

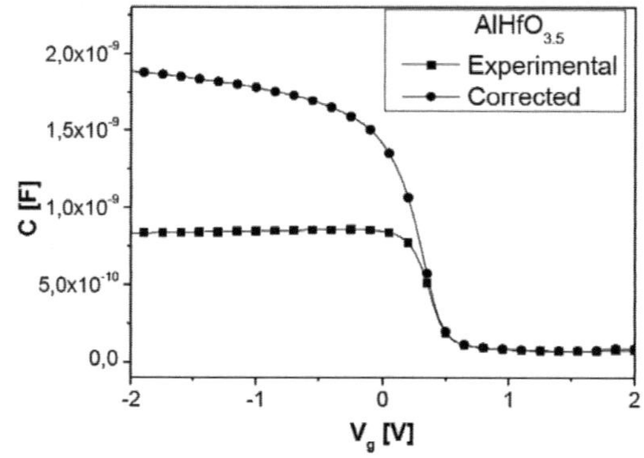

Figure 4. Typical experimental and corrected C-V curves for $AlHfO_{3.5}$.

Figure 4 shows the typical experimental and corrected C-V curves for $AlHfO_{3.5}$, where the correction of the capacitance was performed with the aid of the Equation (4). At first, it is important to highlight that the C-V experimental curve in figure 4 shows a slight decrease of the measured accumulation capacitance when the gate voltage decreases, which indicates an intense leakage process occurring [9-11].

Table 1. C_o, C_{ma}, G_{ma}, N_A, V_{FB} E Q_{SS}/q extracted from C-V characteristic for $AlHfO_{3.5}$

	A1	A2	A3	A4	Average value
C_o (pF)	1883.6	7044.2	4493.8	5144.7	4600±1500
C_{ma} (pF)	836	934	1100	1120	998±110
G_{ma} (μS)	5880	15010	12140	13340	11600±2800
N_A (cm^{-3})	8.4E+13	3.9E+14	7.8E+14	6.8E+14	(4.8±2.5)E+14
V_{FB} (V)	0.25	0.40	0.05	-0.01	0.17±0.15
Q_{SS}/q (cm^2)	-5.9E+11	-2.6E+12	-1.2E+12	-1.3E+12	-(1.4±0.6)E+12

Table 1 shows an average accumulation capacitance (C_{ma}) almost three times lower compared to the corrected one, which means an intense leakage process occurring for $AlHfO_{3.5}$. Equation (5) allows one to appreciate this qualitative observation, this is to say, the higher the G_{ma} parameter, the higher the leakage conductance (Y_C). This fact can be correlated to the GIXRD analysis since $AlHfO_{3.5}$ presented phase separation and crystallization, which is a well-known morphological appearance associated with higher leakage current density as reported in literature [20].

Table 1 also shows the obtained effective charge (Q_{SS}/q) and the flatband voltage (V_{FB}) for the $AlHfO_{3.5}$ film. It was observed positive flatband voltage and negative Q_{SS}/q parameters indicating a negative charging effect in the dielectrics. As a result, $AlHfO_{3.5}$ is a possible candidate to be used as passivating material for power devices because of two important semi-insulating properties: high flatband voltage due to the charging effect and a intense leakage process [3].

IOP Conf. Series: Materials Science and Engineering **76** (2015) 012008 doi:10.1088/1757-899X/76/1/012008

5. Conclusions

The hafnium aluminate was electrically and physically characterized in order to investigate its semi-insulating characteristics as passivating layer for power devices. The deposited films were annealed in pure nitrogen to simulate the thermal budget during a conventional CMOS processing. Phase separation and cristallization took place, promoting the formation of HfO_2 and $Al_{2.4}$ $O_{3.6}$ crystalline phases. Also, non-homogeneous electronic density distribution in the dielectrics gave rise to the formation of several regions which can be viewed as fitting sub-layers, each of them with arbitrary electron density to obtain the real thickness using the PARRAT32 code. In addition, it was observed an average accumulation capacitance (C_{ma}) almost three times lower compared to the corrected one, which means an intense leakage process occurring for $AlHfO_{3.5}$. Also, a positive flatband voltage and a negative Q_{SS}/q parameter were observed indicating a negative charging effect in the dielectrics.

Acnowledgment
The authors would like to thank CNPq for the financial support.

References

[1] Rajab S M , Oliveira I C , M. Massi, Maciel H S, dos Santos Filho S G, Mansano R D 2006 *Thin Solid Films* **515** 170
[2] Oliveira I C, Massi M, dos Santos Filho S G, Ottani C, Maciel H S, Mansano R D 2001 *Diamond and Related Materials* **10** 317
[3] Liss B, Engstrom O 1995 *J. of Appl. Phys* **78** 1824
[4] Yang K J, Hu M 1992 *IEEE Trans. Electron Dev.* **46** 1500
[5] Zhu W J 2001 *Proc. Int. Symp* p 212 *Proc. Int. Symp. VLSI Technology, Systems, and Applications* (Hsinchu, Taiwan, April 2001) (VLSI-TSA) p 212
[6] Ramanathan R, Park M, McIntyre P E 2002 *J. Appl. Phys* **91** 4521
[7] Yu H Y 2002 *Appl. Phys. Lett.* 81 **81** 3618
[8] Stemmer S 2003 *Appl. Phys. Lett.* **83** 3141
[9] Lo S H, Buchanan D A, Taur Y, Wang W 1997 *IEEE Electron Dev. Lett.* **18** 209–11
[10] Lee S W 1994 *IEEE Trans. Electron Dev.* **41** 403–12
[11] Huang C L, Faricelli J V, Arora N D 1993 *IEEE Trans. Electron Dev.* **40** 1134–39
[12] Kuo C S 2004 *IEEE Trans. on Electron Dev.* **51** 854
[13] Bhatta U M, Ghatak J, Mukhopadhyay M, Conley R, Liu C, Satyam P V 2008 *Nuclear Instruments and Methods in Physics Research Section B: Beam Interactions with Materials and Atoms* **266** 1548
[14] Wyon C, Gonchond J P, Delille D, Michallet A, Royer J C, Kwakman L, Marthon S 2006 *Appl. Surface Science* **253** 21
[15] Pietsch U, Holy V, Baumbach T 2004 *High-Resolution X-ray Scattering: From thin Films to Lateral Nanostructures* (Berlin: Springer) pp 143-178
[16] Gibaud A, Vignaud G 2009 *X-ray and Neutron Reflectivity: Specular Reflectivity from Smooth and Rough Surfaces* ed J Daillant A Gibaud (Berlin: Springer)
[17] Terada S, Mrakami H, Nishihagi K, Osaka H 2001 Thickness and density measurement for new materials with combined X-ray technique *Proceedings of the Advanced Semiconductor Manufacturing Conference* (Munich, Germany, 23-24 April 2001) (IEEE/SEMI) pp 125-130
[18] Parratt L G 1954 *Physical Review* **95** 359-69.
[19] Névot L, Croce P 1980 *Revue de Physique Appliquée* **15** 761-79
[20] dos Santos Filho S G, Sonnenberg V, Hora W G, da Silva D M, Kassab L R P 2011 *Thin Solid Films* **520** 2695.

IOP Conf. Series: Materials Science and Engineering **76** (2015) 012009 doi:10.1088/1757-899X/76/1/012009

A new hybrid electrochromic material: vanadium oxide/ eriochrome black T

L F R Junior[1], R S de Oliveira[1], E A Ponzio[1*]
[1] Institute of Chemistry, Department of Physical Chemistry, Group of Electrochemical and Electroanalytical (G_2E), Federal Fluminense University, Niterói, RJ, BRA.

E-mail: [*]eaponzio@vm.uff.br

Abstract. In this work, we describe the synthesis of a new hybrid material, which is consists of the eriochrome black T (sodium 1-[(1-hydroxynaphthcalen-2-yl-hydrazinylidene]-6-nitro-2-naphthol-4-sulfonate) and vanadium pentoxide xerogel using a melting sonoquenching technique. This hybrid material was apply in electrochromic electrodes. The hybrid material was characterized by X-ray difraction, FT-IR, AFM and spectroelectrochemically. The hybrid presented good reversibility and cyclability during 50 cycles studied, electrochromic efficiencies were about 50 $cm^2\ C^{-1}$ (375 nm and 400 nm) and the ΔOD at 375 nm and 400 nm were 0.236 and 0.159 respectively wavelenghts. The response times were 0.7 to 3.3 s for 375 nm and 0.7 to 2.0 s for 400 nm. The hybrid is a potential material to be used in electrochromic applications.

1. Introduction

Vanadium oxide has been studied due its several applications such as secondary batteries and as electrochromic material [1-3]. Among them the electrochromic property draws attention and stands out since it can change its colour due an external application of potential.

Studies show that yellow vanadium oxides films can be converted into green or blue when applying a potential. This colour change is related to a insertion/extracting of ions and electrons from the V_2O_5 structure, creating species with the $M_xV_2O_5$ formula (where M^+ represents ions, such as Li^+ and H^+), and the response of these films are related to some parameters as its crystallinity, degree of disorder and morphology [4-6].

Vanadium oxide has many possible structures, and among them it has a layered structured, known as V_2O_5 xerogel, that allows the intercalation of other materials. Therefore, V_2O_5 xerogel can be classified as an inorganic matrix in which other kinds of materials as organic molecules can be inserted, resulting in an organic-inorganic hybrid material [4,7,8].The new material is interesting because it may have different properties compared with its inorganic matrix, V_2O_5 in this case, resulting in a new electrochromic material. These new materials can have different stages of colour, better response times and good transmittance variations on the UV-Vis spectra when applying a potential.

For example, studies showed that the hybrid $V_2O_5.0.07H_2O.(C_{26}H_{22}N_2O_4Br_2)_{0.26}$ can be synthesized by melting sonoquenching technique [4]. Intercalation of carboxybenzylaviologen was increased the interlayer distance of de vanadium oxide matrix in 0.7 Å. It could also be seen that it has a great transmittance variation (92.7%) and it showed good response times, between 6 and 9 seconds, concluding that the hybrid synthesized has a great potential to be applied into electrochromic electrodes.

IOP Conf. Series: Materials Science and Engineering **76** (2015) 012009 doi:10.1088/1757-899X/76/1/012009

Another nanostructured hybrid was synthesized by melting sonoquenching technique recently [9], $V_2O_5.1.26H_2O([C_3N_2(C_6H_7N)_3])_{0.07}$ nanofibers. The organic guest provokes an interlayer spacing decrease of the inorganic matrix and the electrostatic interaction between the oxo groups of the V_2O_5 matrix and the pyridinium rings of the organic guest demonstrate a strong interaction. The new hybrid nanostructure presented good reversibility and cyclability during 50 cycles, electrochromic efficiency of 22 $cm^2 C^{-1}$ (410 nm) and 96% color retention after 50 cycles of color changing.

In this paper we report the synthesis, using a melting sonoquenching technique, and characterization of an organic-inorganic hybrid, using as inorganic matrix, vanadium oxide as an intercalation material, eriochrome black T (sodium 1-[(1 hydroxynaphthcalen-2-yl-hydrazinylidene]-6-nitro-2-naphthol-4-sulfonate) with potential application into electrochromic electrodes.

2. Experimental Procedure

2.1. Synthesis of hybrid material

The powder of V_2O_5 was first put into alumina crucible, then it was calcined in a muffle furnace at 800°C for one hour and the molten compound (V_2O_5 (l)) was fast poured into water-based saturated solution of the eriochrome black T (sodium 1-[(1-hydroxynaphthcalen-2-yl-hydrazinylidene]-6-nitro-2-naphthol-4-sulfonate) into an ultrasonic bath.

The mixture was left in ultrasonic bath (Unique model USC1400, 40 kHz and 120 W) for 30 minutes. The wet gels were aged for 7 days at room temperature in sealed test tubes. After aging, the sample was washed several times with ethanol until the upper layer became colorless.

The hybrid material was used to deposit thin films on FTO (fluorine-doped tin oxide coated glass substrate) using spin-coating technique. The spin-coater used is homemade equipment. All parameters (rpm, time, concentration of suspension and applied volume) were the same for all material studies to maintain the same geometric area, uniformity surface and film thickness. For all the depositions, 1,000 rpm, 1 min. and 10 drops of a suspension of the materials (50 mg mL^{-1} in isopropyl alcohol) were used. The exposed area was 0.84 cm^2 and the resulting electrodes were dried up at 110 °C for 24 h.

An aliquot of the material was dried up at 110 °C for 24 h and the powder of the hybrid material was characterized by X-ray diffraction, FT-IR, AFM and spectroelectrochemically.

2.2. Measurements

X-ray powder diffraction (XRD) data were recorded on a Bruker-AXS D8 Advanced diffractometer using a LynxEye detector and Cu-Ka emission lines (1.541 Å, 40 kV, 40 mA). The diffraction patterns were collected in a flat geometry and data were collected at room temperature over the range $3 \leq 2\theta \leq 85$ with steps of 0.02° and accumulation time of 0.5 s per step.

Fourier transform infrared spectroscopy (FT-IR) measurements were taken in a Varian 660-FTIR spectrometer.

Atomic force microscopy (AFM) was taken in a Nanosurf Flex AFM. The hybrid material was deposited on a Si substrate and images were acquired by tapping mode at room temperature conditions, using a Nanosurf easyScan 2 Flex AFM scanner and cantilever type Tap190Al-G (BudgetSensors).

Spectroelectrochemical measurements were taken using a potentiostat/galvanostat Microautolab Type III and a spectrophotometer Ocean Optics USB650UV. The electrochemical cell consisted of a quartz cuvette with 1 cm optical path containing three electrodes. The working electrode was glass covered with FTO with the hybrid material deposited on a conducting surface using the technique of spin-coating, the counter electrode was a Pt wire and the quasi-reference electrode was a silver wire coated with AgCl. A solution of 0.5 mol L^{-1} $LiClO_4$ in acetonitrile was used as electrolyte. The optical and electrochemical measurements were carried out at room temperature.

3. Results and Discussion

IOP Conf. Series: Materials Science and Engineering **76** (2015) 012009 doi:10.1088/1757-899X/76/1/012009

XRD measurement of the hybrid material was performed, and is shown in figure 1(a). The XRD pattern indicated the presence of the peak at $2\theta = 7.0°$ for the (001) diffraction plane, as expected for a layered material with parallelism between layers and the reflection plane, corresponding to the interlayer distance of 14.2 Å. It is also observed that in addition to an amorphous phase, there is a crystalline phase of vanadium oxide, it was indexed with V_2O_5 (Shcherbinaite) #96-901-2221, space group Pmn21(31) at orthorhombic crystal system. Another form of characterization was through FT-IR spectroscopy. Figure 1(b) shows the FT-IR spectra for the vanadium oxide in (i), for the eriochrome black T in (ii) and for the hybrid material in (iii). FTIR spectrum of the hybrid shows the presence of some bands that may be related to the organic intercalant present as the band 3406 cm^{-1} related to νOH (which can also be related to the presence of water) and a band at 1621 cm^{-1} relating to the C=C bond of the ring overlying the band νH-O-H of water present in the lamellar inorganic matrix. In addition, it can be seen that there are two bands of low intensity in range of 1600-1000 cm^{-1} (one at 1403 cm^{-1} and the other at 1098 cm^{-1}) possibly related to νC=N in the structure and the νC-C in the ring, respectively. The low intensity of the bands described above is due possibly to the small amount of organic material in the intercalated inorganic hybrid matrix. Moreover, one can notice the presence of the bands at 1006 cm^{-1} (peeled 13 cm^{-1} for wavenumbers when compared to V_2O_5) for the ν V=O which, as can be seen by the spectrum had intensity increased compared with the spectrum inorganic matrix, a fact which is possibly overlapping of this band with the band S=O of the intercalant. Also observed a band at 531 cm^{-1} (10 cm^{-1} shifted to longer wavelengths compared to V_2O_5) can be attributed to ν_{sym}V-O-V. It can also be noted that there is a band at 809 cm^{-1} which is due to δV-O-V, which was taken off 4 cm^{-1} to longer wavelengths when compared with V_2O_5. The displacement of these bands is possibly related to hydrogen bonds present in the hybrid in question, amending the interactions of the pyramidal structure of the matrix.

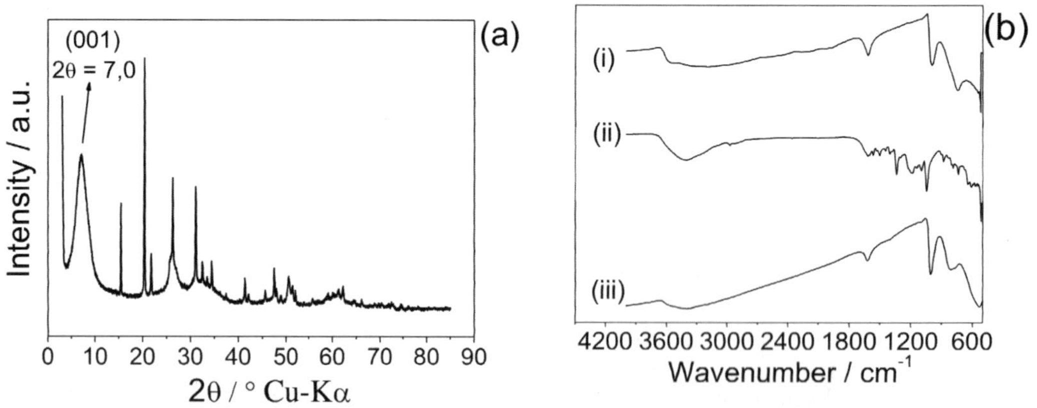

Figure 1. (a) X-ray difraction of the hybrid material. (b) FT-IR measurements for V_2O_5 in (i), eriochrome black T in (ii) and hybrid material in (iii).

AFM measurements were performed in order to analyze the morphology of the synthesized material. It could be seen that the hybrid has a fibrous structure, as it is shown in figure 2. This structure has approximately 15 nm in width and between 0.4 and 0.6 μm in length.

IOP Conf. Series: Materials Science and Engineering **76** (2015) 012009 doi:10.1088/1757-899X/76/1/012009

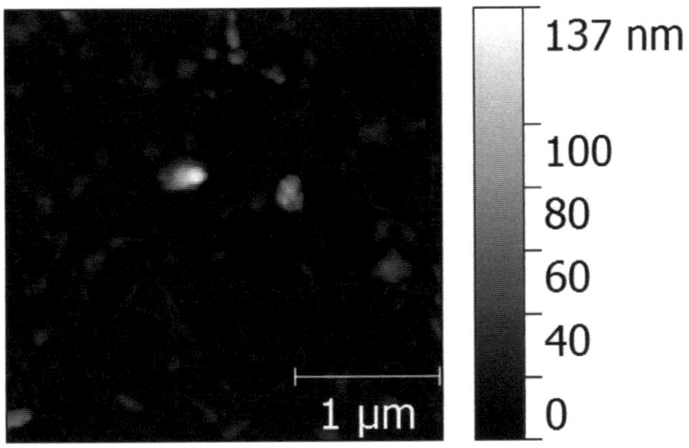

Figure 2. AFM image obtained for the hybrid synthesized.

The spectroelectrochemical characterization of the hybrid synthesized was performed in order to verify its potential to be applied into electrochromic electrodes. Figure 3(a) shows the cyclic voltammogram of the hybrid in a scan rate of 5 mV s^{-1} in the potential range of -0.6 to 1.2 V (vs Ag|AgCl). Figure 3(a) shows a reversible cyclic voltammogram that exhibits three oxidation peaks at 0.18 V, and 0.65 V (possibly related to the oxidation process of V^{4+} to V^{5+}), and at 0.35 V (related to the organic molecule intercalated). Also, three reduction peaks at 0.38V, -0.22V (associated to the reduction of V^{5+} to V^{4+}), and at 0.045 V (possibly associated with a reduction that occurs at the organic molecule).

The absorbance spectra of the hybrid under different applied potentials are shown in figure 3(b). The spectra obtained is different from the spectra observed to the V_2O_5 pure, indicating the synthesis of a new electrochromic material. It can be noted that the UV-Vis spectra showed absorption bands at 327 nm, which is related to π-π* transition, bands at 378 and 462 nm, which are related to charger transfer and a low band at 621 nm related to d-d transitions. It can be noted that during the oxidation process a significant increase in the intensity of absorption of the bands at 327; 378 and 462 nm.

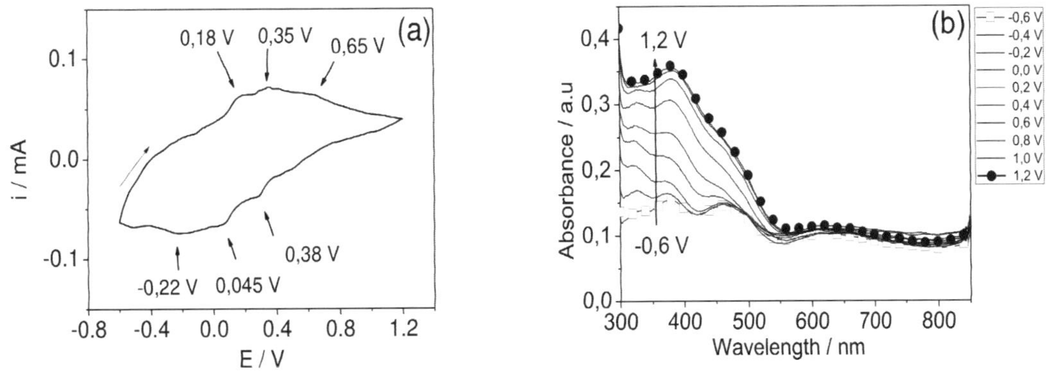

Figure 3. Cyclic voltammogram of the hybrid at 5 mV s^{-1} in (a) and UV-Vis spectra on chronoamperometry at constant current in (b).

Chronoamperometric analysis (not showing here) was performed by monitoring the optical density variation, at the wavelength 375 and 400 nm, which consisted of applying -0.6 V for 30 seconds and, then, 1.2 V for the same time. A total of 50 cycles were conducted. It was possible to note that the transmittance variation (ΔT%) *in situ*, between the oxidized and reduced states, was 79.5% for 375 nm and 77.4% for 400 nm after 50 cycles, respectively.

It was also possible to calculate the response time. In this work, the response time was defined as the required time to reach 75% of the total of the transmittance variation. The results showed that at 375 nm, the response time was between 0.7 and 3.3 seconds, whereas at 400 nm, it ranged between 0.7 and 2.0 seconds. In addition, other parameters were analysed, such as the electrochromic efficiency, which was about 50 cm^2 C^{-1} for both the studied wavelengths and the ΔOD (0.236 at 375 nm and 0.159 at 400 nm).

4. Conclusions

This contribution demonstrates that it is possible to obtain an unprecedented hybrid nanomaterial using the melting sonoquenching techniques. This hybrid, with defined morphology, can be used as electroactive material in electrochromic electrodes. The electrochromic efficiencies were about 50 cm^2 C^{-1} (375 nm and 400 nm) and the ΔOD at 375 nm and 400 nm were 0.236 and 0.159, respectively. The response time was 0.7 to 3.3 s for 375 nm and 0.7 to 2.0 s for 400 nm, respectively. From the measurements of FT-IR and XRD, it was possible to conclude that the hybrid vanadium oxide and the eriochrome black T were successfully synthesized, and AFM measurements showed that it has a fibrous structure. In addition, from the spectroelectrochemical measurement, it has different properties from vanadium oxide and that it could be seen that this hybrid is a potential material to be used in electrochromic applications.

5. Acknowledgments

The authors are grateful to the Coordenação de Aperfeiçoamento de Pessoal de Nível Superior (CAPES), Fundação Carlos Chagas de Amparo a Pesquisa do Estado do Rio de Janeiro (FAPERJ, E-26/102.971/2012 and E-26/111.407/2013), and Pró-reitoria de Pesquisa, Pós-graduação e Inovação (Proppi-UFF) for constant financial support and to LDRX-UFF for the XRD measurements.

References

[1] Cao Z and Wei B 2013 *Nano Energy* **2** 481.
[2] Ren X, Shi C, Zhang P, Jiang Y, Liu J and Zhang Q 2012 *Mater. Sci. Eng. B* **177** 929.
[3] Cheng K -C, Chen F -R and Kai J -J 2006 *Sol. Energy Mater. Sol. Cells* **90** 1156.
[4] Oliveira R S, Alves W A and Ponzio E A 2012 *ECS Trans.* **43** 363.
[5] Oliveira, M R S, Mello D A A, Oliveira R S and Ponzio E A 2011 *Adv. in Nanotechnol.* **8** 1.
[6] Scarminio J, Talledo A, Andersson A A, Passerini S and Decker F 1993 *Electrochim. Acta* **38** 1637.
[7] Oliveira, R S, Oliveira M R S, Oliveira S C and Ponzio E A 2013 *Rev. Virtual Quim.* **5** 596.
[8] Arashiro E, Zampronio E C, Brunello C A, Lassali T A F, Oliveira H P and Graeff C F O 2001 *Int. J. Inorg. Mater.* **3** 727.
[9] de Oliveira R S, Goulart J S, Miranda F S and Ponzio E A 2014 *J. Braz. Chem. Soc.* **25** 540.

IOP Conf. Series: Materials Science and Engineering **76** (2015) 012010 doi:10.1088/1757-899X/76/1/012010

Physical Characterization of Cu-Ni-P Thin Films aiming at Cu/Cu-Ni-P Thermocouples

F. Tomachevski[1], M. Sparvoli[1], S.G. dos Santos Filho[1]

[1]University of São Paulo: LSI/PSI/EPUSP – São Paulo, Brazil

E-mail: fehally@gmail.com, marinsparvoli@yahoo.com.br, sgsantos@usp.br

Abstract. Cu-Ni-P thin films have a high-thermoelectric power, which allows the fabrication of very sensitive heat-flux sensors based on planar technology. In this work, (100) silicon surfaces were pre-activated in a diluted hydrofluoric acid solution containing $PdCl_2$. Following, Cu-Ni-P thin films were chemically deposited using an alkaline chemical bath containing 15 g/l $NiSO_4.6H_2O$; 0.2 g/l $CuSO_4.5H_2O$; 15 g/l $Na_2HPO_2.H_2O$ and 60 g/l $Na_3C_6H5O_7.2H_2O$ at temperature of 80 °C where NH_4OH was added until pH was 8.0. It was noteworthy that the stoichiometric percentages of Ni and Cu vary substantially for immersion times in the range of 1 to 3 min and they become almost stable at 50% and 35%, respectively, when the immersion time is higher than 3 min. In addition, the percentage of P remains almost constant around 17-18 % for all the immersion times studied. On the other hand, the sheet resistance also varies substantially for immersion times in the range of 1 to 3 min. Based on the surface morphology, smaller grains with size in the range of 0.02 to 0.1 μm are initially grown on the silicon surface and exposed regions of silicon without deposits are also observed for immersion times in the range of 1 to 3min. Therefore, the discontinuities and non uniformities of the films are promoting, respectively, the observed behaviours of sheet resistance and stoichiometry.

Introduction

The construction of thermoelectric sensors using thin films of copper and constantan (Cu and Ni in a roughly equivalent ratio) is highly regarded because of the high energy that produce these alloys [1-3]. Starting from the same principle, we can construct these thermocouples on a silicon substrate through a process known as electrochemical process.

Constantan alloys (CuNi) have a very small thermal dependence (almost zero) on its electrical resistivity, which can be noticed by an almost constancy of its thermal coefficient of resistivity ($\alpha = 1x10^{-5}$ K^{-1}) in the range of 20 to 600°C. This dependency also makes agreement with the almost constant coefficient of linear expansion ($\alpha = 13.5x10^{-6}$ -$16x10^{-6}$ K^{-1}, for temperature ranging from 100 to 600°C, which is almost exclusively due to the fact that this alloy has a high level of crystalline structure, in the mentioned temperature range. Thanks to the small temperature coefficients and low electrical resistivity of Constantan, it ended up becoming one of the main materials for manufacturing resistors and precision measurements, as well as heating devices [3].

Unlike the already known process of electroplating, where we have a voltage drop between the electrodes of different metals, in the chemical deposition, the reaction occurs because of the particular conditions of pH on the solution, and the temperature at which it is depositing the metal ions present in the solution (in the case of this work, ions of Cu, Ni and P) on the silicon wafer (Si) [4,5].

In the chemical deposition process, anodic reactions are observed (where there is oxidation reactions) and cathodic reactions (where there is reduction reactions) happening at the same time giving rise to a conductive electrolyte solution [5].

This paper proposes the study of the aforementioned alloy deposition through electrochemical process where will be extracted the composition and concentrations from Rutherford Backscattering analysis, sheet resistance of the material from four-probe measurements and the visualization of the surface morphology, besides demonstrating the benefits of the electrochemical process for thin film deposition.

Experimental

Si wafers were chemically cleaned by RCA cleaning [6]. The recipe used for the RCA cleaning consisted of two baths: (a) 400 ml H_2O + 25 ml NH_4OH (38%) + 175 ml H_2O_2(37%) and (b) 400 ml H_2O + 100 ml HCl (38%), heated at 90°C, for 15 min. Between each bath, the wafers were immersed in deionized (DI) water for 5 min. Following, Palladium pre-activation were performed immerging samples in an acid solution (0.1 g/l of $PdCl_2$ and 50 ml/l of HF) for 60 seconds in order to obtain Pd sites sparsely distributed onto the silicon surfaces. Finally, CuNi alloys were chemically deposited onto n-type Si substrates using an aqueous citrate electrolyte containing Cu and Ni sulfates. The solution used was 7.5 g / l $NiSO_4.6H_2O$, 0.2 g / l $CuSO_4.5H_2O$, 7.5 g / l $Na_2HPO_2.H_2O$, 30 g / l $Na_3C_6H_5O_7.2H_2O$, NH_4OH (pH ≈ 8.0) at 80 °C and thin films were deposited for time immersions of 1, 2, 3, 4, 5, 7 and 10 min.

RBS spectra were taken at 3.0 MeV under normal incidence of a $^4He^+$ beam and with a scattering angle of 170° using an accelerator Tanden, model 55 DH/NEC. In addition, the SiMRA program [7,8] was employed to adjust a simulated spectrum to the experimental one using fitting of composition and areal concentration (N_P(atoms/cm^2)) where N_P is proportional to the RBS-signal area of the formed composite or element [7]. In addition, the stoichiometric percentage of Ni, Cu and P were obtained by means of the relative values of the areal concentrations.

In addition, the sheet resistance of the films was measured by four-point probe and the surface features of the were observed by Field-Emission Scanning Electron Microscopy (FE-SEM).

Results and Discussion

At first, several preliminary tests were performed with different concentrations of $NiSO_4$, $CuSO_4$ and Na_2HPO_2 at a fixed pH of 8 and at a fixed temperature of 80 °C. The addition of copper influences the electroless reaction due to the difference in redox potential between Ni and Cu [2].

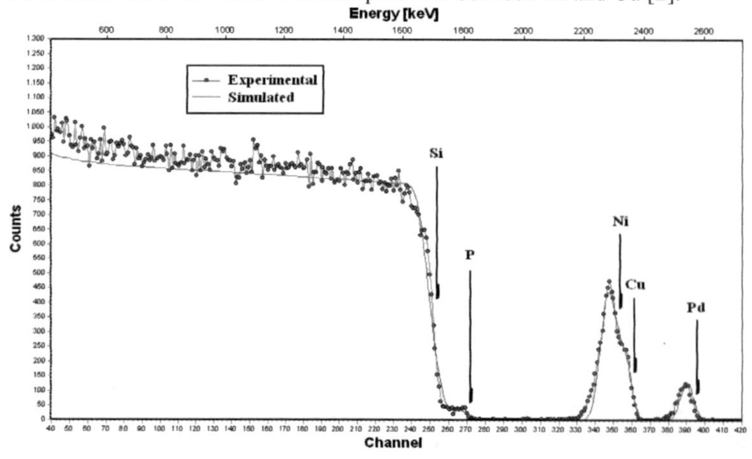

Figure 1. Typical RBS spectrum and simulation for a immersion time of 1min.

Copper is preferentially deposited because of its higher reduction potential than nickel and phosphorus. Thus, it was established that a Ni/Cu atomic ratio of at least 70 in the chemical solution to obtain Ni/Cu atomic ratio of 0.5 in the deposits. After the deposition process, the deposited Cu-Ni-P films were analysed by means RBS analysis, which was interpreted by SIMRA v.6 program. Figure 1 shows the typical RBS spectrum and the simulated curve for a immersion time of 1 min. In this case, the Cu and Ni signals are partially superimposed since the atomic masses are close. However, the SIMRA program allows one to separate both contributions by fitting the signals in order to obtain areal concentration and composition. The adjusted curves are illustrated superimposed on the experimental ones and, as a result, errors lower than 5% were obtained during the extraction of the relative aereal concentrations for Ni, Cu and P.

In addition, figure 2 also shows the typical RBS spectrum and the simulated curve for a immersion time of 4 min where the Cu and Ni signals become almost entirely superimposed since the area under the Cu-Ni signal (aereal concentration) was substantially increased compared to the spectrum shown in figure 1. However, the SIMRA program allowed to separate both contributions by fitting the signals of Cu and Ni.

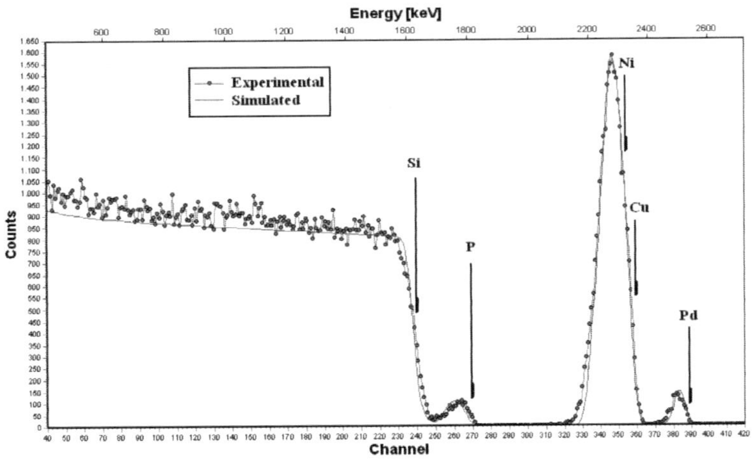

Figure 2. Typical RBS spectrum and simulation for a immersion time of 4min.

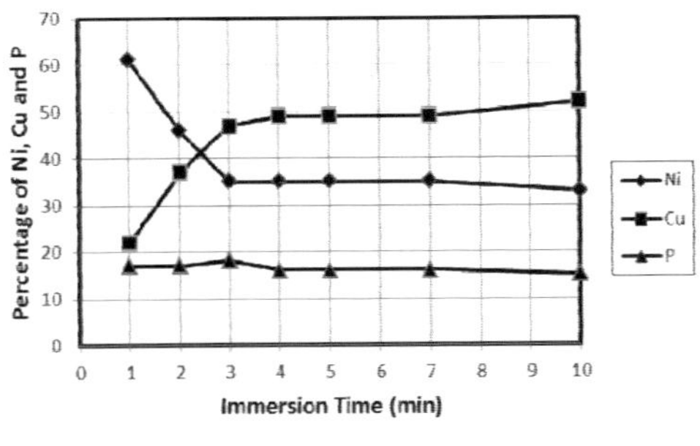

Figure 3. Stoichiometric percentages of Ni, Cu and P as a function of the immersion time.

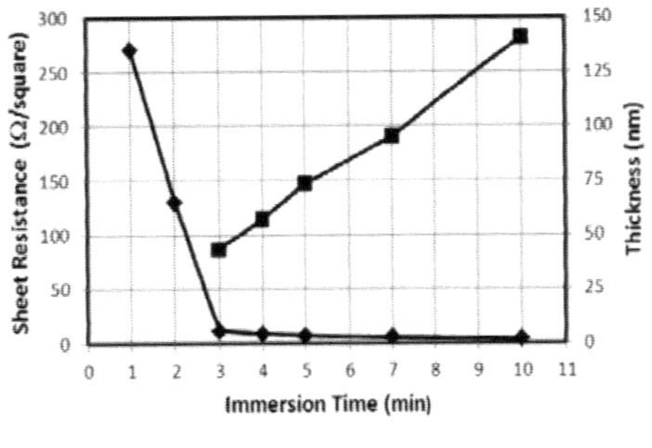

Figure 4. Sheet resistance and thickness as a function of the immersion time.

Figure 3 illustrates the stoichiometric percentage of Ni, Cu and P as a function of the immersion time. It is noteworthy that the stoichiometric percentages of Ni and Cu vary substantially for immersion times in the range of 1 to 3 min and they become almost stable at 50% and 35%, respectively, when the immersion time is higher than 3 min. In addition, the percentage of P remains almost constant around 17-18 % for all the immersion times studied and the sheet resistance also vary substantially for immersion times in the range of 1 to 3 min. Figure 3 suggests a continuous structural change of the morphology for immersion times in the range of 1 to 3 min.

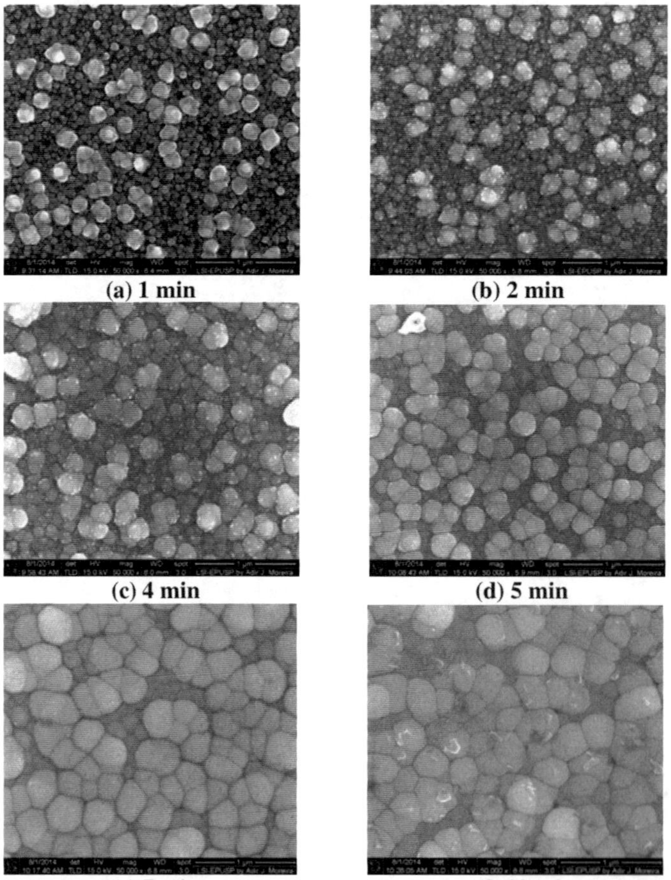

(a) 1 min **(b) 2 min**
(c) 4 min **(d) 5 min**
(e) 7 min **(f) 10 min**

Figure 5. FE-SEM micrographs of the Cu-Ni-P films for immersion times of: (a) 1 min, (b) 2 min, (c) 4 min, (d) 5 min, (e) 7 min and (f) 10 min.

The sheet resistance was measured by means of the four-point probe technique and, the thickness was estimated by the relation between the reference resistivity for constantan (49$\mu\Omega$.cm) [4] and the measured sheet resistance. It was noteworthy that the sheet resistance presents a great change for immersion times in the range of 1 to 3 min, which is closely correlated to the great change of stoichiometry observed in the same range of immersion times. On the other hand, as expected, the sheet resistance is substantially lower for immersion times higher than 3 min. To shed further light on the morphology evolution, the surface features were observed by FE-SEM, as shown in figure 5.

From FE-SEM micrographs presented in figure 5, it is clear that smaller grains with size in the range of 0.02 to 0.1 μm are initially grown on the silicon surface where it is visible exposed regions of silicon without deposits, for immersion times of 1 min and 2 min. Figure 5c already shows a surface without exposed regions for immersion time of 4min (very similar to the one obtained for 3 min, not shown), which means a great change in the morphology as already pointed out by the results of stoichiometry and sheet resistance. In the following, for immersion times higher than 3 min, the average grain size increased from about 0.1 μm to a upper limit of 0.5 μm when the immersion time achieves 10 min. This evolution of the grain size can be correlated with the comcomitant decrease of the sheet resistance from 11.3 to 3.5 Ω/square for the immersion time varying from 3 to 10 min. Therefore, the discontinuities and non uniformities of the films are promoting, respectively, the observed behaviours of sheet resistance and stoichiometry.

1. Conclusions

Cu-Ni-P thin films were chemically deposited using a de-ionized-water-diluted alkaline chemical bath containing 15g/l $NiSO_4.6H_2O$; 0.2 g/l $CuSO_4.5H_2O$; 15 g/l $Na_2HPO_2.H_2O$ and 60 g/l $Na_3C_6H_5O_7.2H_2O$ at a temperature of 80°C where NH4OH was added until pH was 8.0. The concentrations and compositions of the films were obtained by Rutherford Backscattering (RBS) and the surface morphology was characterized by Scanning Electron Microscope (SEM).

Copper preferentially deposited because of its higher reduction potential than nickel and phosphorus. Thus, it was established that a Ni/Cu atomic ratio of at least 70 in the chemical solution to obtain Ni/Cu atomic ratio of 0.5 in the deposits.

It was shown that the stoichiometric percentages of Ni and Cu varied substantially for immersion times in the range of 1 to 3min and they became almost stable at 50% and 35%, respectively, when the immersion time was higher than 3min. In addition, the percentage of P remained almost constant around 17-18% for all the immersion times studied. On the other hand, the sheet resistance also varied substantially for immersion times in the range of 1 to 3min. Based on the surface morphology, smaller grains with size in the range of 0.02 to 0.1μm were initially grown on the silicon surface and exposed regions of silicon without deposits were also observed for immersion times in the range of 1 to 3min. Therefore, the discontinuities and non uniformities of the films promoted, respectively, the observed behaviours of sheet resistance and stoichiometry.

Acnowledgment

The authors would like to thank CNPq for the financial support.

References

[1] Park J J, Taya M 2004 *Electronics Lett.* **40** 599
[2] Delatorre R G Sartorelli M L Schervenski A Q Pasa A A 2003 *J. of Appl. Phys.* **93** 6155
[3] Glatz W, Hierold, C 2007 Flexible micro thermoelectric generator *Proceedings of the IEEE 20th International Conference on Micro Electro Mechanical Systems* (*Kobe, Japan, 21-25 January 2007*) (IEEE) p 89
[4] Mallory G O Hadju J B 1990 *Electroless plating: Fundamentals and Applications* (New York: Noyes Publications)
[5] Byeon J H et al 2008 *Surface and Coating Technology* **202** 3571
[6] Kern W and Puotinen D A 1970 *RCA Rev.* **31** 187
[7] Chu W K, Mayer J W Nicolet M A 1978 *Backscattering Spectrometry* (New York: Academic Press)
[8] Mayer M 1999 SIMNRA, a Simulation Program for the Analysis of NRA, RBS and ERDA *Proceedings of the 15th International Conference on the Application of Accelerators in Research and Industry* (*Denton, Texas, 1999*) ed J L Duggan, I L Morgan (Maryland: American Institute of Physics, vol 475) p 541

IOP Conf. Series: Materials Science and Engineering **76** (2015) 012011 doi:10.1088/1757-899X/76/1/012011

Effect of monolayer supports on the electronic structure of single-layer MoS_2

Alfredo Ramirez-Torres, Duy Le and Talat S. Rahman

Department of Physics, University of Central Florida, Orlando, FL, 32816

Email: Duy.Le@ucf.edu

Abstract. We present results of density functional theory based calculations of the electronic structure of a single-layer of MoS_2, as modified by three different single-layer materials used as support: hexagonal boron nitride (h-BN), graphene, and silicene, using the local density approximation (LDA), on the one hand, and a functional that explicitly includes van der Waals interactions (optB88-vdW), on the other hand. Because the lattice mismatch between the primitive cell of MoS_2 and those of the supports is large, calculations are performed with as large a supercell as computationally feasible, so as to reduce the incommensurability between lattices. Even though van der Waals interactions are expected to play an important role in the binding between MoS_2, and h-BN and graphene, we find that the band structure and related conclusions obtained by optB88-vdW and LDA are quite similar for the three heterostructures considered here. Single-layer MoS_2 interacts weakly with h-BN and graphene, while covalent bonds are formed with silicene. Detailed analysis of the electronic density of states also indicates little effect of h-BN and graphene, while silicene severely modifies it by introducing additional states within the band gap. Furthermore, adsorption on graphene brings the conduction bands of MoS_2 down to the Fermi level of graphene as a result of charge transfer from graphene to MoS_2, while adsorption on silicene shifts both valence and conduction bands towards the Fermi level, in addition to inducing a gap of ~50 meV in silicene itself.

1. Introduction

The experimental production of graphene on 2004 [1] has sparked tremendous interest in the study of two-dimensional (2D) atomic crystals. Layered materials in which the bonding within layers is covalent and that between them is of the van der Waals (vdW) type are a rich source of 2D materials. One set of such materials is the transition-metal dichalcogenide, of which molybdenum disulfide (MoS_2) is a prototype. When MoS_2 is reduced from bulk to a single layer its electronic properties change drastically. Both forms have been experimentally and theoretically shown to be semiconductors, but while in the bulk the gap is indirect (~1.3 eV), in the single layer the optical and fundamental gap are direct (~1.8 eV [2-4] and 2.15 eV [5], respectively) making MoS_2 attractive for electronic and optical applications. Figure 1a shows the structural features of a single layer of MoS_2: this material is formed by an atomic plane of molybdenum sandwiched between two atomic planes of sulfur. Even though single-layer MoS_2 has an thickness of three atomic layers, researchers have successfully found its usage in fabricating electronic devices such as field-effect transistor [6, 7], phototransistors [8].

Evidently, the properties of MoS_2 are affected by the substrate on which it is. For example, h-BN is a good choice as support in order to get higher photoluminescence quantum yield in a single layer of

MoS_2 [9]. Also, the efficiency and properties of MoS_2-based field-effect transistors vary depending upon the substrates used in device fabrications [6, 10]. Understanding the effect of support on properties of MoS_2 is thus important for MoS_2-based device designing and manufacturing. It would be also helpful to locate supports whose interactions with an adsorbed MoS_2 film are either negligible (with virtually no effect on the intrinsic properties of the film) or tunable in ways that produce desired effects. We are interested in the effect of three 2D materials, h-BN, graphene, and silicene, as supports on electronic properties of single-layer MoS_2 because these 2D material substrates share with MoS_2 a hexagonal lattice and because such heterostructures have been experimentally realized [9].

Hexagonal boron nitride has proven to be an attractive dielectric substrate for improving the mobility of charge carriers and stability in graphene [11]. Among its features are its large band gap (~5.97 eV) [12] and its relative lack of dangling bonds, which reduces charge traps. Graphene, on the other hand, is a zero-gap material in which charge carriers near Dirac points behave as massless particles [13]. At the same time, silicene the analogue of graphene for silicon, has been predicted to be stable with a low buckled-honeycomb geometry rather a planar structure owing to the strong tendency of Si atoms to adopt sp^3 hybridization [14]. Just as graphene, silicene is a zero-gap material whose conduction and valence bands meet at the K point of the 2D Brillouin zone.

While we were motivated by the work of Ma et al.[15] who used density functional theory (DFT) to show that for the graphene/MoS_2 heterostructure a gap opens in graphene, we have become aware of a few other related studies in the course of preparation of this manuscript. Komsa and Krasheninnikov [16] using DFT and a semi-empirical correction for van der Waals interactions through the DFT-D2 approach [17]show that for the bilayer heterostructures MoS_2/BN and MoS_2/graphene, the electronic structure of MoS_2 is not modified. The MoS_2/BN heterostructure was found to be a semiconductor while MoS_2/graphene was a semimetal, with the Dirac point of graphene slightly below of the conduction bands of MoS_2. Unlike Ma *et al.* [15], they did not report any gap opening in graphene. Li et al. [18], on the other hand, report a gap opening of the graphene/MoS_2 superlattice which varies with the stacking of the two materials, and that the graphene/MoS_2 superlattice turns metallic and conductions bands of MoS_2 are aligned with the Fermi level. Likewise, a number of other authors [19-22] have carried out calculations for one or more of the heterostructures considered here and their results depend very much of the amount of strain in the system and the choice of the functional used for the DFT calculations.

Our interest in this work is two-fold: 1) carry out a comparative study of the changes in the electronic structure of single-layer MoS_2 brought about by each of three single-layer materials used as a substrate, when the lattice match is small; 2) compare the results obtained using LDA with those from more accurate treatment of van der Waals interactions (as in the code optB88-vdW). It is well known that the LDA overestimates bonding. The general gradient approximation (GGA) designed to correct this deviation, has had reasonable success. On the other hand, both functionals fail to take into account dispersion forces. However, unlike GGA which gives only repulsive interactions in weakly bond systems, LDA frequently gives reasonable results [23-25]. As we shall see, LDA, perhaps fortuitously, provides reasonably good description of the three systems in question here. While it would have been most attractive to have considered systems sizes in which the MoS_2 layer had zero strain, the results presented here have very small (< 1.6% for MoS_2/BN and MoS_2/silicene and < 3% for MoS_2/graphene) lattice mismatch.

2. Computational procedure

Our calculations are based on DFT as implemented in the Vienna *ab initio* Simulation Package (VASP) [26, 27], and Quantum Espresso package (QE) [28]. For VASP calculations, we use projector augmented wave (PAW) [29, 30] pseudopotentials generated using Perdew-Burke-Ernzerhof parameterization (PBE) [31] and plane-wave basis set (with a cut-off energy of 500 eV) methods. We utilize the optB88-vdW [32, 33] version of the van der Waals density functional (vdW-DF) [34, 35] because of its success in the treatment of dispersion bonded systems [33]. For simplification in the text and for emphasizing the explicit inclusion of vdW interaction, we refer herein opbtB88-vdW as vdW.

For QE calculations, the exchange and correlation effects are treated within the local density approximation (LDA), using the Perdew-Zunger (PZ) functional [36]. Electronic wave-functions are expanded using a plane-wave basis set with a kinetic energy cutoff of 60 Ry. We used norm-conserving pseudo-potentials [37].

The lattice mismatch between primitive cell of MoS_2 and the supports here considered is large (experimental lattice parameters of MoS_2, h-BN, and silicene are 3.16 Å [38], 2.504 Å [39], and 2.46 Å [40], respectively). Previous theoretical studies have shown that the use of strong or free strained heterostructures lead to different results [15, 16, 20]. In order to minimize the changes in the electronic structure coming from the artificial strain introduced to get commensurability between lattices, we have considered, for each MoS_2/substrate combination, supercells with large periodicities in which the lattice mismatch is small and is computationally feasible for DFT calculations. Additionally, because we consider the single-layer MoS_2 as the adsorbate, in all cases its lattice parameter will be slightly adjusted to get commensurability with the single-layer supports, while the lattice parameters of the supports will be fixed to their optimized values. Parameters of supercells were generated using our calculated lattice constant of support materials and will be described more in details in the next section. A vacuum of ~15 Å was used to separate normal periodical images to minimize the interaction between them. Models were fully relaxed until all components of force acting on each ion are smaller than 0.01 eV/Å. The reciprocal space was sampled with a 6×6×1 mesh generated through Monkhorst-Pack scheme [41] for MoS_2/graphene and MoS_2/BN, while with a 3×3×1 mesh for MoS_2/silicene.

3. Structural properties

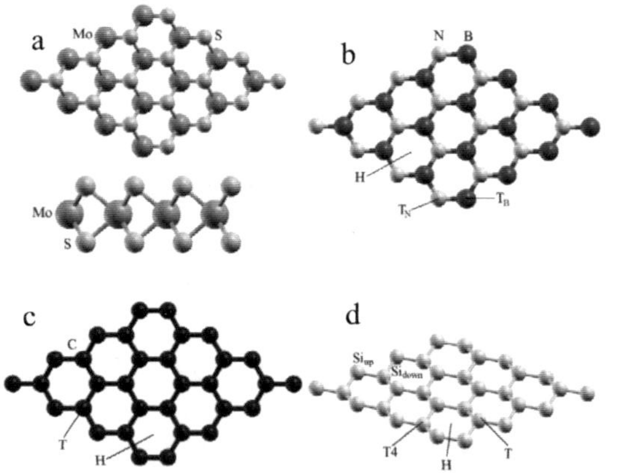

Figure 1. Structural representations of (a) single-layer MoS_2, (b) single-layer boron nitride, (c) graphene, and (d) silicene. For the supports (b-d), the high-symmetry adsorptions sites are labeled as H (hole), T4 (tetrahedral), and T (top).

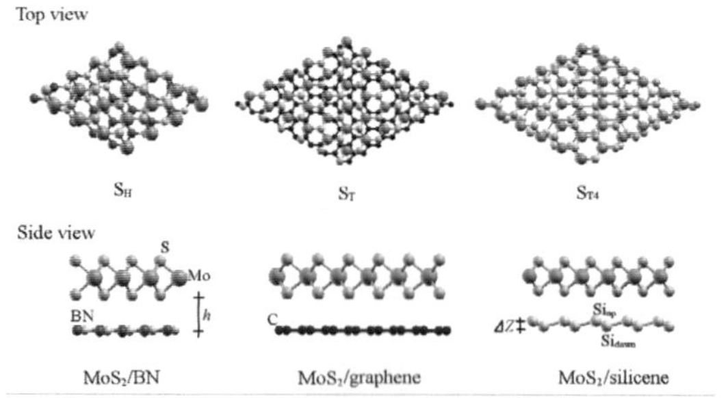

Figure 2. Relaxed models to represent the adsorption of a single layer of MoS_2 on different supports. h in the figure represents the distance between layers, and ΔZ is the height difference between Si atoms.

We found a lattice parameter of 3.187 Å (3.167 Å) with the vdW (LDA) functional for single-layer MoS_2 which is less than 1% difference from experimental value [38], of 2.508 Å (2.502 Å) for h-BN, of 2.465 Å (2.448 Å) for graphene, and of 3.863 Å (3.81 Å) silicene. We considered in this work the system consisting of (4×4) MoS_2 and (5×5)h-BN, (3×3) MoS_2 and (4×4)graphene, (6×6) MoS_2 and (5×5) silicene with lattice mismatches of -1.65% (-1.32%), 3.03% (2.91%), and 0.99% (0.18%), respectively. The LDA always gives a lower lattice mismatch. Here, we defined the lattice mismatch as $(1-(na/n_0a_0))$; where (n_0, a_0) and (n, a) are the (number of periodical units, lattice parameters) of the support and of the adsorbed MoS_2 film, respectively. Positive (negative) lattice mismatch indicates that the adsorbed film is under tensile (compressive) strain. To find the optimal adsorption structures of a single-layer MoS_2 on the support in question, we constructed several models on the basis of the periodicities listed above. To fit the single-layer MoS_2 over the support we choose a sulfur atom or a molybdenum atom belonging to the single-layer MoS_2 as an initial point of reference, and place it on a high-symmetry site of the support (see Figure 1. b, c, and d) allowing the single-layer MoS_2 and the support to have the same orientation. We repeated this process for each possible combination of S or Mo atom and types of high-symmetry sites. It is important to note that we adjusted the MoS_2 lattice parameter to make it commensurable with the support.

We obtained three non-equivalent models for MoS_2 on h-BN (S_{TB}, S_{TN}, and S_H, those are equivalent with Mo_{TN}, Mo_H, and Mo_{TB}, respectively), two on graphene (S_T and Mo_T, those are equivalent with S_H and Mo_H, respectively), and two on silicene (S_T and Mo_T, those are equivalent with S_H, S_{T4} and Mo_H, Mo_{T4}, respectively). Herein, Name of each configuration is simplified by X_Y, where X (= S or Mo) is the reference atom of MoS_2, Y is the high-symmetry site of support on top of which X is placed. See figure 1 for explanations of Y subscripts. For the same support, the difference of the total energy between non-equivalent models is very small, of the order of only tenths of meV resulting in that we can regard these conformations of the three systems as isoenergetic. It is worth mentioning that the total energy of vdW-heterostructures are generally insensitive to the stacking order between layers [15, 18, 19, 21]. Nerveless, we chose to study in details the configurations with the lowest energy. For the case of vdW functional, they are S on hole (S_H) for MoS_2/BN, S on top (S_T) for MoS_2/graphene, and S on top of a tetrahedral site (S_{T4}) for MoS_2/silicene, as presented in Figure 2. For the case of LDA functional, they are and S on top of B (S_{TB}) for MoS_2/BN, S on top (S_T) for MoS_2/graphene, and Mo on top of a tetrahedral site (Mo_{T4}) for MoS_2/silicene. We find the average vertical distance (h) between MoS_2 and support is always lower in LDA approximation. The structural parameters of the heterostructures are summarized in Table 1.

Table 1. Structural parameters of models used to represent the adsorption of a MoS_2 single layer on 2D substrates. All distances are in ångstroms.

Length (Å)	**MoS_2/BN**		**MoS_2/graphene**		**MoS_2/silicene**	
	vdW	**LDA**	**vdW**	**LDA**	**vdW**	**LDA**
Mo-Mo	3.136	3.128	3.286	3.264	3.219	3.174
Mo-S	2.406	2.412	2.440	2.439	2.425	2.423
MoS_2/support (h)	3.353	3.279	3.373	3.318	3.205,3.742	2.882,3.406
B-N	1.448	1.445	-	-	-	-
C-C	-	-	1.423	1.413	-	-
Si_{up}-Si_{down}	-	-	-	-	2.294	2.261
ΔZ	-	-	-	-	0.50-0.57	0.43-0.60

In order to compare the relative stability of our models configurations we calculated the binding energy (BE) per unit area of MoS_2 for each system, i.e., the difference between the energy of the relaxed heterostructure and that of relaxed single-layer components, divided by the number of unit areas of MoS_2, as summarized in the equation $BE = [E_t - (E_s + E_i)]/n$, where E_t, E_s, and E_i are the total energy of, respectively, the system, the support, and the isolated MoS_2 layer prior to adsorption, and n is the number of unit cells of adsorbed MoS_2 layer in the considered supercell. Negative values of BE indicate adsorption of MoS_2 on the substrate. For the considered heterostructures we find BE (calculated using vdW functional) to be -160 meV per unit-cell of MoS_2 for MoS_2/BN, -120 meV for MoS_2/graphene, and -190 meV for MoS_2/silicene. The respective values calculated with LDA functional are -77, -79, and -142 meV. Both vdW and LDA approximations indicate that MoS_2 binds most strongly on silicene. Although the vertical distance between MoS_2 and support is always lower in LDA, the absolute value of binding energy is always larger in vdW case.

4. Electronic structure of heterostructures

Our calculated band structures for the three heterostructures considered here, together with that of simple layer MoS_2 (for comparison) are presented in Figure 3 for vdW case and in Figure 4 for LDA case. In agreement with previous findings [15, 20], single-layer MoS_2 is a direct band gap semiconductor, with a band gap ~1.67 eV for vdW and ~1.81 eV for LDA (this latter one coincides with the experimental optical gap) at the K point. It is important to mention that the calculated band gaps reported here are much smaller than that calculated using GW method [42] (~2.75 eV [43]) and that measured experimentally (~2.15 eV [5]).

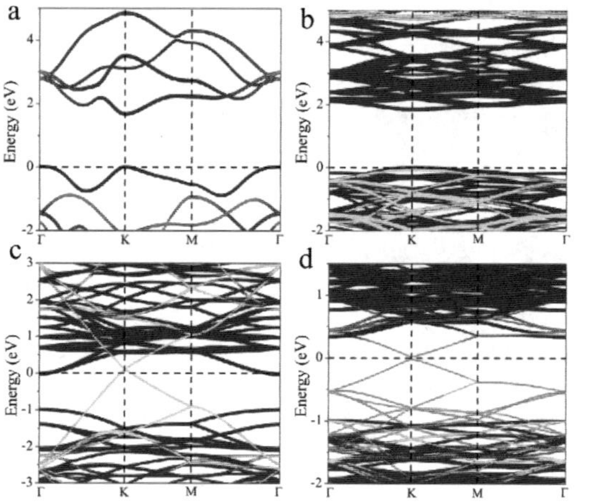

Figure 3. Calculated band structure in vdW case of (a) isolated single-layer MoS_2 (Blue dots correspond to Mo states, while magenta dots are sulfur states), (b) MoS_2-(4×4)/ BN-(5×5), (c) MoS_2-(3×3)/ graphene-(4×4), and (d) MoS_2-(6×6)/ silicene-(5×5). In (b), (c), and (d), blue and green dots show the contribution from MoS_2, and from support materials, respectively.

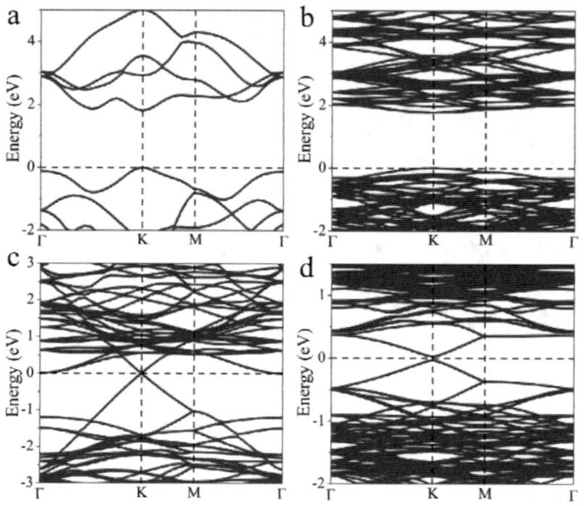

Figure 4. Calculated band structure in LDA case of (a) isolated single-layer MoS_2, (b) MoS_2-(4×4)/ BN-(5×5), (c) MoS_2-(3×3)/ graphene-(4×4), and (d) MoS_2-(6×6)/ silicene-(5×5).

In order to understand the band structure of these systems it is useful to recall that when the periodicity of a supercell increases, the first Brillouin zone of the supercell diminishes in size, so that bands of the normal cell are folded into a smaller Brillouin zone. The first Brillouin zones of all films treated here have hexagonal geometry which allows us adopt the 3N-rule [44] to comprehend how *k*-points in the extended Brillouin zones can be folded into the primitive one: for (3N±1) × (3N±1) periodicities, K points of the primitive cell are folded to K points of the supercell's Brillouin zone; for (3N) × (3N) periodicities, K points of the primitive cell are folded to Γ point of the supercell.

Calculated band structure of the MoS_2/BN heterostructure with the vdW approximation (see Figure 3b) shows a direct gap of approximately 1.86 eV, 0.19 eV larger than that of an isolate single-layer MoS_2 (figure 3a) while that with LDA functional also indicates a direct band gap that is slightly (0.03 eV) larger than that of a single layer (see figure 4a, 4b). The difference in the band gap values can be traced to the artificial strain effect [45]. Apart from this, the band structures of each single layer components do not exhibit significant changes from its isolate ones.

Figure 3c and 4c show band structures of MoS_2 on graphene system calculated using vdW and LDA functionals, respectively. Since graphene is a zero-gap material, it introduces states inside the gap of the single-layer MoS_2 and the heterostructure is a semimetal. From these band structures calculated using vdW functional (figure 3c), it is clear that graphene is *p*-doped and MoS_2 is *n*-doped because the Dirac point of graphene aligns above the Fermi level while the bottom of conduction bands (BCB) of MoS_2 has occupation suggesting a charge transfer from graphene to MoS_2. In LDA case (figure 4c), the effect is not that pronounced. However, BCB of MoS_2 and the Dirac point of graphene are also alignment to Fermi level indicating a much smaller charge transfer than that in the case of vdW functional. It is also worth mentioning that results from both functionals do not show any gap-opening in graphene, in contradiction with results reported by Ma et al. [15], suggesting that the interaction between MoS_2 and graphene is not strong enough to break symmetry of graphene.

Figures 3d and 4d show band structures of MoS_2 on silicene system calculated using vdW and LDA functionals, respectively. As in the case of MoS_2/graphene, silicene introduces states inside the MoS_2 gap. Interestingly, there is a small gap opening at the Dirac point of silicene and the entire system turns out to be a semiconductor with a small gap of ~50 meV for vdW, and 55 meV for LDA. The gap opening indicates that the interaction between MoS_2 and silicene is strong enough to break symmetry of silicene similar to what has been reported in by Li et al. [21]. This result is in good agreement with that from Gao et al. [46]. However, they found a gap of 70 meV for silicene/MoS_2 system. The difference can be mainly traced to the functional (DFT-D2) and the lattice mismatch of 2.42% and PBE-D functional used by Gao et al. [46] resulting in the stronger interaction between MoS_2 and silicene. The parameter ΔZ (which indicates the difference in height between Si_{up} and Si_{down}, see Figure 2) was 0.50 Å in vdW, and 0.44 Å in LDA for our relaxed cell of isolate silicene, while in relaxed MoS_2/silicene heterostructure ΔZ has several values distributed in the range 0.50-0.57 Å in vdW and 0.43-0.60 in LDA (as shown in the Table 1). These values imply that the LDA-calculated interactions between MoS_2 and silicene causes a stronger modification to silicene than vdW-calculated one does attesting once again that the silicene opening is larger in the LDA case.

Further insights into the band structure may be obtained from the electronic density of states (DOS), with information about the contribution of the orbitals of each species, through the projected density of states (PDOS). Because PDOS obtained by both approximations, vdW and LDA, are very similar, we only present the vdW results. Figure 5a shows the DOS of an isolated single-layer MoS_2: states on both sides of and near the band gap mainly have Mo-*d* character, and a strong orbital mixing between S-*p* and Mo-*d* states can be appreciated (we have used our 4x4 supercell to compare with the MoS_2/BN system explicitly).

Figure 5b shows total DOS and various partial contributions (only orbitals with major contributions) for MoS_2/BN system. Not surprisingly, the shape of the DOS of an isolated single-layer MoS_2 (figure 5a) and of a MoS_2/BN heterostructure (figure 5b) are somewhat similar (in the range of energy shown). Additionally the top valence band (TVB) of h-BN is almost aligned with the TVB of MoS_2. Hence no electronic state is introduced in the gap of the single-layer MoS_2 and no occupied states of the BN appear until around of 4.7 eV. Comparing PDOS of S and Mo atoms in the single-layer MoS_2 and those in the MoS_2/BN system, we can appreciate their similarities, including the width of band gap and the majority Mo-*d* character in the top of the valence bands and bottom of the conduction bands, indicating that the BN does not cause any significant effect on the electronic structure of MoS_2.

Figure 5c shows the electronic DOS of MoS_2/graphene system. The PDOS confirms that BCB of the single-layer MoS_2 are shifted towards the Fermi level of graphene (see figure 5c). The states inside of the MoS_2 gap come mainly from C-*p* orbitals. The V-shape in the DOS of graphene remains intact after MoS_2 adsorption (see the inset in figure 5c). The shape of the total DOS above the Fermi level is not surprisingly determined mainly by Mo-*d* orbitals. The artificial strain applied to single-layer MoS_2 when conjoined with graphene causes the valence-band maximum of MoS_2 to move to the Γ point; thus MoS_2 behaves as an indirect semiconductor, with TVB and BCB in Γ and K points respectively giving a gap of ~1 eV in vdW, and ~1.2 eV in LDA. The alignment of conductions bands of MoS_2 to Fermi level of graphene can be exploited for fabrication of phototransistors [47] or photovoltaic devices [48], in which electron-hole pairs are produced in MoS_2 by absorption of light, and electrons move to graphene, while holes are trapped in MoS_2.

Examination of the PDOS in MoS_2/silicene system (Figure 5d) reveals that MoS_2 and Si states extend towards the middle of the original gap of the isolated single-layer MoS_2. At around -1.9, -1.6 and -1.1 eV it is possible observe a slight orbital mixing (hybridization) between MoS_2 and silicene, consequently we expect not only vdW interactions between layers in this system but also covalent bonds. The shape of the total DOS near the Fermi level is dominated by Si-*p* orbitals. However, beyond this region the shape of total DOS is determinate mainly by Mo-*d* orbitals.

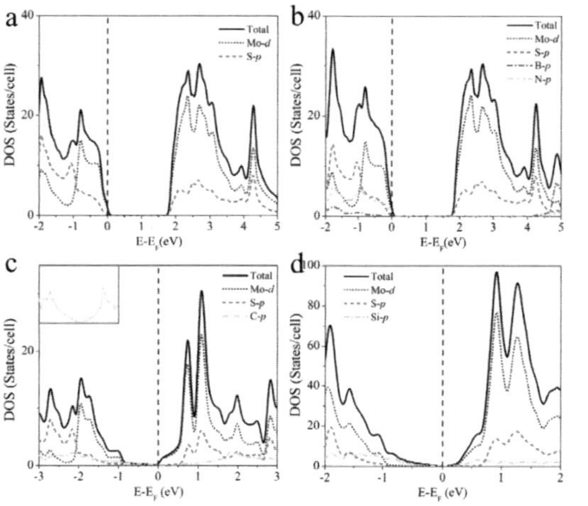

Figure 5. Electronic density of states for (a) an isolated single-layer MoS_2, (b) MoS_2/BN, (c) MoS_2/graphene, and (d) MoS_2/silicene.

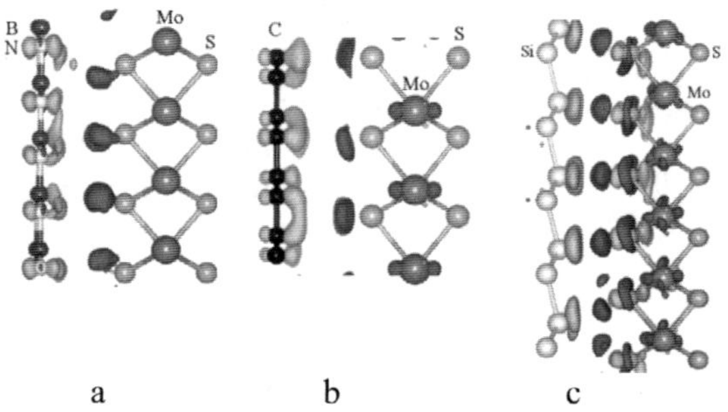

Figure 6. Plots of 3D electron density difference ($\Delta\rho$), for (a) MoS_2/BN at the isosurface value of 2×10^{-4} e/Å^3, (b) MoS_2/graphene at the isosurface value of 3×10^{-4} e/Å^3, (c) MoS_2/silicene at the isosurface value of 5×10^{-4} e/Å^3. The red (green) areas represent electron density increase (decrease).

Bader-charge analysis [49-51] for vdW case indicates no charge transfer from h-BN to MoS_2 or vice versa and about 0.01 and 0.04 electrons from graphene and silicene, respectively, to an unit cell of a single-layer MoS_2. Löwdin-charge analysis [52] in LDA case indicates that there is no charge transfer from graphene or BN to MoS_2 and 0.02 electrons from silicene to MoS_2 causing it n-doped. Heterostructure with silicene shows the greatest charge transfer. In order to investigate the density redistribution in our systems, and to confirm the nature of bonding between the single layers we calculated the electron density difference $\Delta\rho = \rho_T - \rho_{Sup} - \rho_{MoS_2}$; where ρ_T, ρ_{Sup}, ρ_{MoS_2} are the electron density of the heterostructure, the support in question and the single layer of MoS_2, respectively. Figure 6 shows the 3D electron density difference plots for each system at different isosurface values. In MoS_2/BN and MoS_2/graphene heterostructures charge redistribution is observed. However, according to Bader and Löwdin-charge analysis, the charge transfer is very small, thus in these systems vdW interactions are dominant. In MoS_2/silicene significant density rearrangement in MoS_2 layer occurs. Furthermore there is an accumulation of charge in the interlayer region suggesting orbital overlap between layers. These results confirm that unlike the vdW-dominance in MoS_2/BN and MoS_2/graphene systems, in MoS_2/silicene system besides the vdW interactions weak covalent interactions exist.

5. Conclusions

Results of DFT calculations of the structural and electronic properties of a single-layer MoS_2 supported on single-layer of: h-BN, graphene or silicene show the DFT-vdW (optB88-vdW) and LDA approximations for the XC functional to lead to very similar conclusions. For the both functionals we find the total energy to be insensitive to the stacking order between layers. Of the three considered supports, h-BN is an excellent option to preserve the electronic structure of single-layer MoS_2 because it leaves the band structure virtually the same as that of an isolated single-layer. On the other hand, graphene and silicene modify the electronic structure of MoS_2 by introducing states into its band gap and causing it to become n-doped. We find that in MoS_2/BN, single-layer MoS_2 has a direct band gap whose value is somewhat modified because of the artificial strain in the system. The MoS_2/graphene heterostructure is a semimetal, where bottom of conduction bands of MoS_2 is shifted to the Fermi level. The expansion applied on MoS_2 to match with graphene results in the indirect gap of MoS_2. In MoS_2/silicene, silicene introduces states inside the gap of isolated MoS_2 and MoS_2 states themselves are extended towards the middle of its original gap. The interaction between MoS_2 and silicene opens a gap in silicene. Thus, the heterostructure has a direct band gap of ~0.05 eV at K point. Though small, ideally this gap is sufficient to prevent electronic transitions at low temperature. The analysis of binding energy, charge transfer, and electron density difference indicate that the interaction between a single-layer MoS_2 and silicene includes weak covalent interactions and is stronger than that on BN or graphene.

Acknowledgements
This work was partially supported by DOE grant DE-FG02-07ER46354. A.R.T. would like to acknowledge CONACYT (Mexico) for a support through the Postdoctoral Fellowship Program (number 204065). The authors acknowledge the University of Central Florida Stokes Advanced Research Computing Center for providing computational resources.

References

[1] Novoselov K S, Geim A K, Morozov S V, Jiang D, Zhang Y, Dubonos S V, Grigorieva I V and Firsov A A 2004 *Science* **306** 666-9
[2] Ellis J K, Lucero M J and Scuseria G E 2011 *Applied Physics Letters* **99** 261908-3
[3] Mak K F, Lee C, Hone J, Shan J and Heinz T F 2010 *Physical Review Letters* **105** 136805
[4] Splendiani A, Sun L, Zhang Y, Li T, Kim J, Chim C-Y, Galli G and Wang F 2010 *Nano Letters* **10** 1271-5
[5] Zhang C, Johnson A, Hsu C L, Li L J and Shih C K 2014 *Nano Letters* **14** 2443-7
[6] Radisavljevic B, Radenovic A, Brivio J, Giacometti V and Kis A 2011 *Nature nanotechnology* **6** 147–50
[7] Zhang Y J, Ye J T, Matsuhashi Y and Iwasa Y 2012 *Nano Letters* **12** 1136–40
[8] Yin Z, Li H, Li H, Jiang L, Shi Y, Sun Y, Lu G, Zhang Q, Chen X and Zhang H 2012 *ACS Nano* **6** 74–80
[9] Mak K F, He K, Shan J and Heinz T F 2012 *Nat Nano* **7** 494-8
[10] Das S, Chen H-Y, Penumatcha A V and Appenzeller J 2012 *Nano Letters* **13** 100-5
[11] Dean C R, Young A F, MericI, LeeC, WangL, SorgenfreiS, WatanabeK, TaniguchiT, KimP, Shepard K L and HoneJ 2010 *Nat Nano* **5** 722-6
[12] Watanabe K, Taniguchi T and Kanda H 2004 *Nat Mater* **3** 404-9
[13] Novoselov K S, Geim A K, Morozov S V, Jiang D, Katsnelson M I, Grigorieva I V, Dubonos S V and Firsov A A 2005 *Nature* **438** 197-200
[14] Cahangirov S, Topsakal M, Aktürk E, Şahin H and Ciraci S 2009 *Physical Review Letters* **102** 236804
[15] Ma Y, Dai Y, Guo M, Niu C and Huang B 2011 *Nanoscale* **3** 3883-7
[16] Komsa H-P and Krasheninnikov A V 2013 *Physical Review B* **88** 085318
[17] Grimme S 2006 *Journal of Computational Chemistry* **27** 1787-99
[18] Li X D, Yu S, Wu S Q, Wen Y H, Zhou S and Zhu Z Z 2013 *The Journal of Physical Chemistry C* **117** 15347-53
[19] Lu N, Guo H, Wang L, Wu X and Zeng X C 2014 *Nanoscale* **6** 4566-71
[20] Zongyu H, Chaoyu H, Xiang Q, Hong Y, Wenliang L, Xiaolin W, Xiangyang P and Jianxin Z 2014 *Journal of Physics D: Applied Physics* **47** 075301
[21] Li X, Wu S, Zhou S and Zhu Z 2014 *Nanoscale Research Letters* **9** 110

IOP Conf. Series: Materials Science and Engineering **76** (2015) 012011 doi:10.1088/1757-899X/76/1/012011

[22] Gao N, Li J C and Jiang Q 2014 *Physical Chemistry Chemical Physics* **16** 11673-8
[23] De-Li C, Al-Saidi W A and Johnson J K 2012 *Journal of Physics: Condensed Matter* **24** 424211
[24] Vanin M, Mortensen J J, Kelkkanen A K, Garcia-Lastra J M, Thygesen K S and Jacobsen K W 2010 *Physical Review B* **81** 081408
[25] Le D, Kara A, Schroder E, Hyldgaard P and Rahman T S 2012 *Journal of Physics: Condensed Matter* **24** 424210
[26] Kresse G and Hafner J 1993 *Physical Review B* **47** 558–61
[27] Kresse G and Furthmüller J 1996 *Comput. Mat. Sci* **6** 15-50
[28] Giannozzi P e a 2009 *Journal of Physics: Condensed Matter* **21** 19
[29] Kresse G and Joubert D 1999 *Physical Review B* **59** 1758–75
[30] Blöchl P E 1994 *Physical Review B* **50** 17953–79
[31] Perdew J P, Burke K and Ernzerhof M 1996 *Physical Review Letters* **77** 3865–8
[32] Klimeš J, Bowler D R and Michaelides A 2011 *Physical Review B* **83** 195131
[33] Klimeš J, Bowler D R and Michaelides A 2010 *Journal of Physics: Condensed Matter* **22** 022201
[34] Thonhauser T, Cooper V R, Li S, Puzder A, Hyldgaard P and Langreth D C 2007 *Physical Review B* **76** 125112
[35] Dion M, Rydberg H, Schröder E, Langreth D C and Lundqvist B I 2004 *Physical Review Letters* **92** 246401
[36] Perdew J P and Zunger A 1981 *Physical Review B* **23** 5048-79
[37] Troullier N and Martins J L 1991 *Physical Review B* **43** 1993-2006
[38] Young P A 1968 *J. Phys. D: Appl. Phys.* **1**
[39] Lynch R W and Drickamer H G 1966 *The Journal of Chemical Physics* **44** 181-4
[40] Okada Y and Tokumaru Y 1984 *Journal of Applied Physics* **56** 314-20
[41] Monkhorst H J and Pack J D 1976 *Physical Review B* **13** 5188-92
[42] Onida G, Reining L and Rubio A 2002 *Reviews of Modern Physics* **74** 601-59
[43] Cheiwchanchamnangij T and Lambrecht W R L 2012 *Physical Review B* **85** 205302
[44] Ye-Cheng Z, Hao-Li Z and Wei-Qiao D 2013 *Nanotechnology* **24** 225705
[45] Yun W S, Han S W, Hong S C, Kim I G and Lee J D 2012 *Physical Review B* **85** 033305
[46] Gao N, Li J C and Jiang Q 2014 *Physical Chemistry Chemical Physics*
[47] Zhang W, Chuu C-P, Huang J-K, Chen C-H, Tsai M-L, Chang Y-H, Liang C-T, He Jr H, Chou M-Y and Li L-J 2014 *Scientific Reports* **4** 3826
[48] Bernardi M, Palummo M and Grossman J C 2013 *Nano Letters*
[49] Henkelman G, Arnaldsson A and Jónsson H 2006 *Computational Materials Science* **36** 354-60
[50] Sanville E, Kenny S D, Smith R and Henkelman G 2007 *Journal of Computational Chemistry* **28** 899-908
[51] Tang W, Sanville E and Henkelman G 2009 *Journal of Physics: Condensed Matter* **21** 084204
[52] Löwdin P O 1950 *The Journal of Chemical Physics* **18** 365-75

AUTHOR INDEX